AF805468

GeoJournal Library

Urban Perspectives from the Global South

Series Editors
Christian M. Rogerson, School of Tourism and Hospitality,
University of Johannesburg, Auckland Park, Gauteng, South Africa
Gustav Visser, Geography and Environmental Studies, Stellenbosch University,
Matieland, South Africa

The Urban Perspectives from the Global South brings together a wide variety of urban scholars under one series title and is purposefully multi-disciplinary. The publications in this series are theoretically informed and explore different facets of varying sized urban places. This series addresses the broad developmental issues of urbanization in developing world countries and provides a distinctive African focus on the subject. It examines a variety of themes relating to urban development in the global South including: city economic development, issues of local governance, urban planning, and the impact of multi-ethnic and multicultural formations in urban affairs. The series aims to extend current international urban debates and offer new insights into the development of urban places in the Global South from a number of disciplines including geography, sociology, political science, economics, as well as urban studies. A special focus of the series is the challenges of urbanization and cities in Africa.

Ronnie Donaldson
Editor

Socio-Spatial Small Town Dynamics in South Africa

Springer

Editor
Ronnie Donaldson
Department of Geography and Environmental Studies
Small Town Research Unit
Stellenbosch University
Stellenbosch, South Africa

ISSN 0924-5499 ISSN 2215-0072 (electronic)
GeoJournal Library
ISSN 2511-2171 ISSN 2511-218X (electronic)
Urban Perspectives from the Global South
ISBN 978-3-031-37141-7 ISBN 978-3-031-37142-4 (eBook)
https://doi.org/10.1007/978-3-031-37142-4

© The Editor(s) (if applicable) and The Author(s), under exclusive license to Springer Nature Switzerland AG 2023
This work is subject to copyright. All rights are solely and exclusively licensed by the Publisher, whether the whole or part of the material is concerned, specifically the rights of translation, reprinting, reuse of illustrations, recitation, broadcasting, reproduction on microfilms or in any other physical way, and transmission or information storage and retrieval, electronic adaptation, computer software, or by similar or dissimilar methodology now known or hereafter developed.
The use of general descriptive names, registered names, trademarks, service marks, etc. in this publication does not imply, even in the absence of a specific statement, that such names are exempt from the relevant protective laws and regulations and therefore free for general use.
The publisher, the authors, and the editors are safe to assume that the advice and information in this book are believed to be true and accurate at the date of publication. Neither the publisher nor the authors or the editors give a warranty, expressed or implied, with respect to the material contained herein or for any errors or omissions that may have been made. The publisher remains neutral with regard to jurisdictional claims in published maps and institutional affiliations.

This Springer imprint is published by the registered company Springer Nature Switzerland AG
The registered company address is: Gewerbestrasse 11, 6330 Cham, Switzerland

Contents

1 A Review of Recent Small Town Research in South Africa 1
Ronnie Donaldson and Musfiqah Majiet

2 The Relationship Between White Privilege, Segregation, Topophobia and Symbolic Boundaries in Secunda, Mpumalanga 17
Tarryn N. K. Paquet

3 Neglect of Municipal Infrastructure in Phalaborwa 47
Manfred Spocter

4 Collaborative Governance, Social Capital and Drought: A Case Study of a Collaborative Governance Regime in Graaff-Reinet 79
Rebecca Light

5 Surviving the Tides: A Study of Small-Scale Fisheries and Community Livelihoods in Selected Coastal Towns of the Western Cape, South Africa 107
Samantha Williams

6 The State and Implications of Housing Provision in Small Towns: Experiences in the Raymond Mhlaba Local Municipality, Eastern Cape Province 127
John Ntema, Robert Mongwe, and Eliot Mathebula

7 The Socio-economic Wellbeing of Small Mining Towns in the Northern Cape 145
Avril Gardiner and Ronnie Donaldson

8 Complex Town-Gown Relationships and the Paradoxes of Studentification in the South African Traditional Authority Lands 169
Nothile Ndimande

9 Local Government Election 2021 Trends in Small Towns 193
Stephen Rule

10 The Historical Development of South African Small Towns as Spa Resorts 225
Christian M. Rogerson and Jayne M. Rogerson

Chapter 1
A Review of Recent Small Town Research in South Africa

Ronnie Donaldson and Musfiqah Majiet

1.1 Introduction

Although small towns have historically been a research lacuna, they are home to almost as many people as secondary cities in the country, and more people live in small towns than in larger towns. There has been a gradual increase in scholarly attention to small town studies in South Africa since the early 2000s. Initially, much of the research focused on local economic development (LED) (e.g., Nel, 2005; Gibb & Nel, 2007; Nel & Rogerson, 2007) and tourism issues, specifically second homes (e.g., Hoogendoorn et al., 2005; Hoogendoorn & Visser, 2010). Since the publication of the edited volume, Small Town Geographies in Africa: Experiences from South Africa and Elsewhere (Donaldson & Marais, 2012), there has been a significant increase in research on small towns in South Africa across several disciplines, including geography, urban planning, sociology, economics, and tourism studies. Most notably, small towns have gained the attention of policymakers, thanks to lobbying efforts by various stakeholders, including academics. However, in their review of small town research, Hoogendoorn and Visser (2016: 95) noted that "the body of literature focused on small towns in South Africa to date [2016] is relatively limited in both scope and scale." To address this gap, Hoogendoorn and Visser (2016) suggested several research themes that they deemed important for further exploration. These themes include the impact of welfare payments on small towns, poor governance, environmental issues, housing concerns, the presence of foreign-owned small businesses (and related xenophobia), and the need for long-term investigations. Hoogendoorn and Nel (2019) conducted the most recent review of the broad literature on small town scholarship in South Africa. Their review

R. Donaldson (✉) · M. Majiet
Department of Geography and Environmental Studies, Small Town Research Unit, Stellenbosch University, Stellenbosch, South Africa
e-mail: rdonaldson@sun.ac.za

© The Author(s), under exclusive license to Springer Nature Switzerland AG 2023

R. Donaldson (ed.), *Socio-Spatial Small Town Dynamics in South Africa*, GeoJournal Library, https://doi.org/10.1007/978-3-031-37142-4_1

found that small towns have started to receive greater academic attention, particularly on themes such as population and economic change, innovation, policy shortfalls, the importance of post-productivist activity, dependency, and depopulation. New challenges are posed by climate change and local government failings, which make the search for development alternatives all the more pressing (Hoogendoorn & Nel, 2019: 153).

In recent years, many small towns in South Africa have experienced decline due to factors such as population loss, economic decline, and a lack of investment. However, there has also been growing interest in revitalizing these towns, with a focus on economic development, tourism, and heritage preservation (Donaldson, 2018). Small towns typically face a range of socioeconomic challenges, including poverty, unemployment, and inequality (Toerien, 2020). Research in this area has explored the root causes of these issues, as well as potential solutions such as job creation, skills training, and social support programs. Urban sprawl is not limited to larger cities and metros; in small towns, this process occurs through formal large-scale gated developments (Spocter, 2016) or informal settlements on the peripheries, resulting in the loss of local identity and declining traditional agricultural-based industries, which severely impact the local environment. However, it is not all doom and gloom. Many small town studies have shown how through various LED, tourism, and community initiatives small towns can prosper and grow (Donaldson, 2018; Campbell, 2016).

In this chapter, we provide a brief review of some of the key research themes that have emerged over the past 10 years. We begin with the two themes that have remained crucial in scholarly debate, namely LED and tourism-related research. We then turn our focus to emerging themes such as infrastructure, the housing dimension, urban development and upgrading, and mining town issues. We conclude the review with recommendations for future research.

1.2 Local Economic Development (LED)

Extensive research has been conducted on local economic development (LED) in small towns, with many studies examining the intersection of these two areas (Toerien & Seaman, 2014; Toerien, 2018; Toerien, 2022; Rogerson, 2016a, b, c; Nel & Rogerson, 2007; Rogerson, 2016a, b, c; Nel, 2017). The success of small towns in terms of economic performance is largely influenced by factors such as municipal management, geographic location, and setting, with coastal towns, mountainous landscapes, aesthetically appealing regions, and historically preserved small towns often experiencing greater success (Toerien, 2018). LED and access to resources are also key factors in the success of small towns (Toerien, 2018).

Agriculture and tourism are the two most important business sectors contributing to economic development in small towns (Toerien, 2018). The emergence of tourism in small towns coincided with the development of a post-productivist rural landscape (Hoogendoorn, 2010), and successful small towns have been able to capitalize on their tourism resources.

Recent research has highlighted the contribution of cultural and creative industries (CCI) to economic development in small towns (Drummond & Snowball, 2019). However, successful development of CCI is dependent on several factors, including the existence of a creative class, a consumer base for CCI goods and services, and an existing tourism industry. Factors such as infrastructure, accessibility, network and collaborating potential, as well as the impact of activities in neighbouring regions, should be further explored to better understand the potential for CCI development (Drummond & Snowball, 2019).

Research conducted by Hoefnagels et al. (2022) explores the local economy of Makhanda using the concept of urban scenes, a situational analysis that is place-specific and contextually based, to plan and implement LED initiatives. The researchers found that tourism and education are the two prominent scenes in Makhanda, with top private schools and Rhodes University driving education-led migration, while creativity and arts, festivals, heritage, wildlife safaris, and hunting contribute to tourism. The town's economic growth is threatened by the local municipality's challenges and its inability to build and maintain successful LED initiatives. To foster an inclusive economy, the authors suggest increasing employment opportunities, improving cooperation between the municipality and stakeholders, and implementing more LED initiatives. These recommendations are relevant for many small towns facing similar challenges.

Local economic development agencies (LEDAs) are vital for managing, directing, and coordinating economic development initiatives in small, peripheral municipalities (Lawrence, 2013; Khambule & Gerwel-Proches, 2019). The success of LEDAs is influenced by three factors: (1) the local economic potential, (2) capacity and resource challenges, and (3) the institutional readiness of both the agency and the parent municipality to manage local development outcomes (Lawrence, 2013). Although LEDAs represent a means for place-based development, interventions have been largely unsuccessful in addressing spatial inequalities (Lawrence & Rogerson, 2019; Lawrence, 2013). LEDA's also require certain measures, as a pre-requisite for it to be carried out successfully in peripheral regions. Although, LEDA's form part of LED in South Africa, they are not fully implemented in the system of local governance. LEDA's often have a particular life cycle that goes hand in hand with local government and the political and administrative context thereof (Lawrence & Rogerson, 2019). In Kwa-Zulu Natal case studies it was found that LEDA's were managed very poorly as the municipalities do not have the necessary tools and programmes put in place to support LEDA's (Khambule & Mtapuri, 2018). To ensure a successful LEDA lifecycle, Nel (2017) proposes five measures, including establishing partnerships with stakeholders, having committed and adequate leadership, and having a clear strategic vision for local development.

Other scholars argue that local businesses and the informal economy also contribute to LED in small towns. Khumalo et al. (2019) suggest that the contribution of the informal economy should be considered in terms of economic development and improving the livelihoods of local traders while creating more employment opportunities in the informal sector. Similarly, in a case study of youth-owned small businesses in the Thulamela local municipality in Limpopo, Kativhu et al. (2021) note the challenges faced by these businesses, including poor infrastructure, competition, and financial inadequacy, and suggest that support initiatives should consider these factors. They further suggest that support initiatives should take into account (and be guided by) the challenges faced by these youth-owned businesses and the context of these areas if they want their initiatives to yield successful support programmes and make a contribution to local economies. Business incubation has proven to be successful in supporting youth-owned businesses and small local businesses (Masutha & Rogerson, 2014). Such initiatives are aimed at providing small enterprise developments with business and technical support in an attempt to grow and develop small enterprises into financially independent enterprises. However, state-supported incubators are more prevalent in lower income areas, such as small towns, whereas private-supported incubators are more common in larger metropolitan areas in South Africa (Masutha & Rogerson, 2014). Moreover, to increase the efficiency and success of business incubators, it should be more localised in lower income areas to support small and local enterprises. Jili et al. (2019) emphasize the importance of community participation in LED projects, arguing that active community involvement, where local resources are utilized and services are tailor-made for locals, is beneficial for local communities. Rogerson and Rogerson (2020a, b) investigates using tourism assets for leveraging LED in King Sabata Dalinyabo in the Eastern Cape, finding that local assets underperformed due to institutional and governance shortcomings, such as corruption and the lack of implementation of development plans. Therefore, there is a need for asset management at the municipal level.

Service delivery poses significant limitations on LED interventions in poorer municipalities, as highlighted by Tsheola and Mokgokong (2012). For instance, a rural municipality would prioritize upgrading infrastructure over enhancing LED initiatives, resulting in rural, under-performing municipalities lagging behind better-performing ones. In the case study of Mthonjeni Municipality, Zulu and Mubangizi (2014) found that poor infrastructure development and limited financial resources, along with an inability to attract private sector investments for LED, hindered the municipality's LED efforts.

According to Rogerson (2016a, b, c), climate change can impact LED initiatives' longevity, particularly in South Africa, which is vulnerable to climate change's effects on both tourism and LED. Therefore, Rogerson (2016a, b, c) suggests that climate change should be considered in LED and tourism support programs to help relevant stakeholders manage tourism assets and maximize tourism and LED opportunities. As an emerging theme in small town scholarship, Hoogendoorn and Nel (2019) argue that developmental challenges like climate change and weak local government impede successful LED initiatives.

Hoogendoorn and Visser (2016) recommend shifting small town research's focus from LED to other themes such as governance, institutional capacity, land and resource access, environmental considerations, food supply, and skills development. Similarly, Houghton et al. (2013) call for a more comprehensive inquiry into

LED in small towns, including social and human capital, micro-level decision-making, and daily relationships between different actors. They contend that considering other facets in LED small town research can lead to a better understanding and, consequently, improved economic development projects in small towns. Toerien (2022) suggests that LED planning should focus on commercial agricultural products and services [primary products sold externally], processors, factories, construction services, and tourism and hospitality services.

1.3 Tourism Development

With the exception of a few studies by specialists in the fields of tourism marketing and economics, the majority of scholarly tourism research conducted in a South African context since the 2000s has been undertaken by geographers (Donaldson, 2018). Six reviews on broad tourism scholarship have illustrated this by identifying specific focus areas in tourism geography (Visser & Hoogendoorn, 2011; Rogerson & Visser, 2014; Hoogendoorn & Rogerson, 2015; Rogerson, 2016a, b, c; Donaldson, 2018, 2021). The scholarly contributions on small town tourism over this period have been grouped into five fields. Firstly, the bulk of research has focused on second homes. Secondly, research into LED and developmental issues of small town tourism has continued to gain strength. Thirdly, investigations into the economic impacts of tourism, especially festivals, have been carried out. Fourthly, nature-based tourism and rural dynamics have come to the fore. Lastly, niche tourism themes have appeared, of which route tourism and visiting friends and relatives (VFR) are the most prominent. In his seminal work on small town tourism in South Africa, Donaldson (2018) provided a comprehensive review of the literature on this theme. His review casts a wider net to incorporate small town tourism scholarship across disciplines in South Africa over a ten-year period. In addition to reviewing the five themes mentioned above, his book also covered some new unexplored themes in South African contexts, such as the dynamics of branding small towns, Africa's first "slow city" (cittaslow), Sedgefield, and a renewed look at rural and small town gentrification with the small town of Greyton as a case study.

Unravelling the shifting trajectory of the tourism space economy remains a major scholarly challenge (Rogerson, 2016a, b, c). Rogerson's (2016a, b, c) study contributed to understanding the changing mosaic of South Africa's tourism space economy in small towns and rural areas. His study has shown that tourism flows in non-metropolitan/secondary cities are strongly dominated by domestic rather than international travellers and that the core purpose of travel for most small towns and rural areas is VFR tourism. He does however highlight differentiation between small towns in the former four provinces of apartheid South Africa and those in the former apartheid-created homelands. The study of Tseane-Gumbi and Ani (2019) specifically focused on tourism and transformation in South Africa's former homelands.

Since the review of Donaldson (2018) a number of new research themes have emerged. We will briefly reflect on four of those here. Firstly, is the underexplored topic of the impact of climate change on small town tourism. In this regard, the pioneering work of Hoogendoorn et al. (2016: 1) claim that climate change "poses considerable threats to low-lying coastal towns including flooding induced by sea-level rise, increased discomfort from changes in temperature and precipitation, more frequent extreme events, biodiversity shifts and water shortages". Their tourism climate index suggests that climate change experienced in recent decades has detrimentally affected tourist comfort in two case study small coastal towns of St Francis Bay and Cape St Francis. Hoogendoorn and Fitchett (2018) in addition provided perspectives on second homes, climate change and tourism by arguing that remote rural municipalities with small towns are often characterised by limited financial capital and thus a weakened adaptive capacity to adapt to the increasing climate change threats such as sea level rise, flooding, drought and storms, which are projected to amplify in severity and frequency under climate change. In Hoogendoorn's (2021) unique exploration of the concept "last chance tourism" (LCT) he argues that predominantly 'tourism towns' like Knysna may become a LCT destination if appropriate fire management measures are not implemented. Secondly, a theme that would probably be explored more in the coming years is that of the devastation caused by the COVID-19 global pandemic. It has been the focus of investigating the impact thereof on rural and small town tourism (Donaldson et al., 2021b; Rogerson & Rogerson, 2022; Rogerson & Sixaba, 2022) as well as on safari lodge industry with its concomitant impacts on nearby towns (Visser & Marais, 2021). Thirdly, the growing body of knowledge on heritage and CCI related studies (Buchanan, 2021; Drummond & Snowball, 2019). Although not a small town, the latent opportunities for heritage tourism in the secondary city of Mahikeng's surrounding rural areas were explored (Drummond et al., 2021) whereas cultural clustering in small towns in a selected Eastern Cape municipal area was investigated (Drummond & Snowball, 2019). Fourthly, there has been a specific focus on tourism accommodation, with a macro analysis and case study-based studies (Rogerson, 2019a, b; Rogerson & Rogerson, 2019a, b).

There have been ongoing studies into economic development and tourism, including research by Munien et al. (2018), Dlomo and Rogerson (2020), and Kontsiwe and Visser (2019). In contrast to earlier pro-poor tourism research, Toerien's (2020, 2022) study confirmed that communities in towns with more tourist and hospitality enterprises are overall wealthier, and vice versa, indicating that tourism is a driver of prosperity and poverty reduction. However, Rogerson and Rogerson's (2019a, b) economic development study found that while tourism development in the Overstrand Municipality contributes significantly to job creation and economic development, it is not inclusive, reflecting the power elite's role in small towns as outlined by Donaldson's (2018) urban growth machine theoretical application. Effective destination marketing remains crucial for the development of tourism in small towns, as demonstrated by Donaldson et al. (2021a) examination of misconceptions surrounding actual destination image creation through marketing in Swellendam.

According to Rogerson (2016a, b, c), taking a macro-level perspective on the function of small towns and rural areas as tourism destinations creates opportunities for future research to delve deeper into comparative studies that can clarify and account for the unique variations in the tourism industry across different localities.

1.4 Housing and Urban Infrastructural Development Challenges

Surprisingly, there are very few studies on housing provision, access, and the role of housing in social and economic development in small towns. Following the end of apartheid and in an effort to rewrite the spatial apartheid patterns, the democratic government implemented various housing programs. However, decades later, housing provision has failed to meet the demand, with little progress in realizing these policies on the ground, resulting in an ever-increasing backlog for state-funded housing.

Inadequate capacity by the local municipality, challenges in accessing suitable land, and poor administration are among the reasons for the lack of housing provision in rural small towns (Ngcayi, 2013). Ntema (2021) argues that the lack of investment in low-cost housing and the inability of rural municipalities to attract private sector investment have hindered the slow delivery, or lack thereof, of state-funded housing. Poor workmanship resulting in cracked walls, leaking roofs, and poorly installed infrastructure leads to a growing backlog as houses become uninhabitable for beneficiaries, forcing local municipalities to terminate projects during the construction phase, and leaving intended beneficiaries back on the waiting list (Manomano & Tanga, 2018). This poor workmanship also forces beneficiaries to spend money on repairs and maintenance, exacerbating an already exhaustive waiting list and housing backlog. For example, in the case study of Napier, those who were on the housing list until 2005 only received housing in 2013 (Oldfield et al., 2020). Without adequate housing provision, there is a direct increase in informal structures and backyard dwellings (Shackleton & Blair, 2013). Furthermore, the lack of housing provision results in a plethora of issues related to essential services, such as water and sanitation delivery (Zama, 2020).

In addition to the growing backlog and increase in informal settlements, the lack of housing provision and incomplete housing projects also adversely affects intended and potential beneficiaries. The most significant impacts include a lack of access to housing and increased poverty. Housing provision and access extend beyond physical infrastructure and provide beneficiaries with a sense of identity, belonging, and citizenship (Social Work Policy Initiative, 2006). Therefore, potential beneficiaries risk losing their worth and sense of identity if not provided with adequate housing (Manomano & Tanga, 2018). Studies also show that beneficiaries' lives have not improved as there is no clear evidence of upward social mobility (Ngcayi, 2013).

The mismatch between government policies and reality suggests the need for a restructuring and revamp of government policies intended for rural development and to increase state-funded projects in small towns (Hebinck et al., 2023). However, research into the role played by non-governmental organizations (NGOs) in assisting the government with housing provision in small towns remains limited. Additionally, there is a lack of scholarship focusing on housing provision in small towns in general.

Infrastructure plays a crucial role in the developmental aspects of human settlements. However, there is a lack of research on infrastructure in small towns in South Africa. Studies mostly rely on case-based analysis, focusing on water, sustainability, and governance issues. Research on Water, Sanitation, and Hygiene (WASH) in the small town of (West Coast) highlights the need for adequate infrastructure to make small towns resilient in the face of natural disasters (Abrams et al., 2021). Despite this, the inability of local municipalities to provide essential services, such as water, has resulted in residents taking on the responsibility of providing these services within their communities. This has become the norm in many small towns, such as Tshakhuma in Limpopo Province (Hofstetter et al., 2021). The study found that most of these self-supply initiatives operate without legal frameworks, indicating the need for institutional structures to collaborate on self-supply initiatives.

Infrastructure for renewable energy, especially for the generation of solar and wind energy, is a relatively new feature in the countryside landscape of small towns. Lombard and Ferreira (2015) explored the growth and spatial distribution of newly commissioned infrastructures for wind and solar energy and the role they can play in the diversification of rural economies in parts of the Northern Cape, Western Cape, and Eastern Cape provinces. However, further exploration is needed to understand the socio-economic spin-offs and impacts on local economies of renewable energy.

In drought-stricken regions of the country, Zhang et al. (2019) propose that managed aquifer recharge (MAR) could be a solution to drought in small towns in the Western Cape province, contributing to water security along the West Coast. Besides infrastructure and municipal interventions, there is a call for research to shift focus to green infrastructure. Shackleton et al. (2018) are pioneers of research focusing on green infrastructure in small towns in South Africa (see also Shackleton & Njwaxu, 2021; Gwedla & Shackleton, 2019; Shackleton & Blair, 2013). Green infrastructure, in the form of public parks and street trees, serves a range of functions from service provisioning to cultural services (Shackleton et al., 2018). Their findings suggest that green infrastructure solutions are more cost-effective than built infrastructure and provide a range of ecosystem services. In a study investigating the demise of urban neighbourhood parks six small towns in the Eastern Cape (Adelaide, Alexandra, Bathurst, Bedford, Kenton-on-Sea and Makhanda) underpins the value of community involvement in small town renewal processes (Shackleton & Njwaxu, 2021). Privatisation of public spaces, such as golf courses, also showed the neoliberal approach to small town growth and management (Spocter, 2017).

The retirement-and-tourism town of Parys serves as a case study that highlights the range of urban environmental issues that not only affect the future growth and development of the town, but also impact public spaces, food security, and service delivery (Dryer, 2013). The lack of essential infrastructural services hinders the ability of local municipalities to position themselves as successful agents of social and economic development. This further deters private investors from investing (Van Rooyen, 2018), resulting in a limited ability for residents of small towns to improve their livelihoods. For example, Khapayi and Celliers (2016) found that a lack of infrastructure, especially roads, in the Eastern Cape hinders the ability of emerging farmers to switch from subsistence to commercial farming. Similarly, Shackleton and Luckert (2015) found that Mbashe and Lukanji local municipalities experienced a decline in agricultural activity due to poor infrastructure and transport systems. In the Western Cape province, residents of an informal settlement in Napier raised concerns related to safety and hygiene due to a lack of infrastructure (Oldfield et al., 2020). In this regard, Dryer (2013) identified the need for small towns to determine the state of sustainability and proposed that they should develop an integrated sustainability database that could assist municipalities.

Scholars have studied the impact of industry in general, and mining activities in particular, on the spatial, social, and economic conditions of communities in South African towns and cities for decades. Examples of such studies include Hanekom (1976), Marais and Cloete (2009), Marais (2013), and Rogerson (2012). The recent surge of research papers on mining towns, with Marais as a prominent author, is a result of various projects conducted on mining towns across the country (Marais, 2023; Marais et al., 2022; Marais et al., 2022; Marais et al., 2021; Burger et al., 2018; Rogerson, 2012). For a more comprehensive overview of mining town scholarship, refer to Chap. 7 in the book. In a similar vein, despite being on South Africa's national development agenda for long now, Saldanha Bay region's industrial expansion has been struggling for years to meet the preconditions for economic take-off (Welman & Ferreira, 2016).

Several developmentally-related research themes have garnered attention, including climate change, community gardens, and the role of development agencies. Simon et al.'s (2021) study provides a first attempt at unpacking the realities of addressing climate change in small-town contexts from a governance perspective. In terms of food security, urban agriculture is believed to play a crucial role in impoverished small-town communities. However, Roberts and Shackleton's (2018) study revealed declines in the number and size of urban community gardens in selected case study towns in the Eastern Cape. The study found that only 16% of the gardens present in the 1980s were still operating in the 2000s. Apart from Ntema's (2021) study on Alice, little is known about small university towns, especially those in the former homeland areas. The role played by development agencies and faith-based organizations remains an underexplored theme (McKibbin et al., 2012). Small town academic scholarship has capacity to inform policy and decision-making, but this is slow to trickle down. Exceptions include studies on the growth potential of small towns in the Western Cape, which were conducted by Donaldson, Du Plessis, et al. (2012a); Donaldson, et al. (2012b); and Van Niekerk et al. (2016). These studies prioritized areas spatially according to their developmental potential and social needs at a municipal level, and their findings were incorporated into the provincial spatial development framework, serving as guidelines for justifying large capital projects in the province.

1.5 Conclusion: Looking Towards to the Future of Small Town Research

This literature review on small town studies in South Africa is not intended to be exhaustive. Although we aimed to concentrate on the geography of small towns, we acknowledge the numerous studies that have been conducted on this topic. Based on our review, we have identified seven themes that merit further exploration in the near future.

The first theme involves the need to consider the histories and traces of small towns in present social dynamics, as emphasized by Mabin (2021). Understanding this can aid in addressing the challenges small towns face. Linked to this theme is that of heritage conservation. The second theme is the need for research on community development as a way of revitalizing small towns. This can involve studying the role of local organizations, community networks, and social capital in promoting economic and social development. The third theme pertains to housing delivery in small towns in South Africa, which presents several challenges, including inadequate resources, outdated infrastructure, and a lack of financial support from government agencies. This area has been underexplored to date. The fourth theme relates to the rise and growth of Zoom towns and gentrification-led small-town development, which require a multifaceted response. This includes policies to protect vulnerable residents, support for affordable housing development, and the promotion of mixed-income communities. Decision-makers should involve the community in developing solutions to ensure that gentrification does not undermine the social, cultural, and economic fabric of small towns in South Africa. The fifth theme is service delivery issues. Despite government efforts to improve service delivery, many small towns still encounter challenges ranging from a lack of basic services like electricity and water to insufficient healthcare, education facilities and social services. The sixth theme is the broader socio-political issue of state and city-capture, including the challenge of corruption, which undermines essential service provision by diverting resources away from their intended use. Civil society organizations and community-based organizations can play a role in raising awareness of corruption and advocating for greater transparency and accountability in service delivery, and this needs to be investigated. Lastly, in 2022 the Department of Cooperative Governance and Traditional Affairs launched the Small Town Regeneration Strategy and Implementation Plan (Department of Cooperative Governance, 2022) to address inadequate infrastructure in small towns. This policy document should be a guideline for future research analysis into its successes.

In the edited volume "Socio-Spatial Small Town Dynamics in South Africa," the editor has compiled a selection of contributions that focus on small town geographical aspects that have not been commonly studied before. Topics include white privilege, municipal infrastructure neglect, collaborative governance, small-scale fisheries livelihoods, housing provision, wellbeing in mining towns, studentification in a rural context, election trends, and the historical development of small-town spas.

References

Abrams, A. L., Carden, K., Teta, C., & Wågsæther, K. (2021). Water, sanitation, and hygiene vulnerability among rural areas and small towns in south Africa: Exploring the role of climate change, marginalization, and inequality. *Water, 13*(20), 2810.

Buchanan, L. (2021). *The status of urban heritage conservation: competency of local government in the Western Cape province.* [Masters thesis]. Stellenbosch University.

Burger, P., Marais, L., & Van Rooyen, D. (Eds.). (2018). *Mining and community in South Africa. From small town to iron town*. Routledge.

Campbell, M. (2016). A great place and not yet another declining small town: The case of Clarens, South Africa. *Urbani izziv, 27*(2), 138–148.

Department of Cooperative Governance. (2022). *Small town regeneration and implementation plan.*

Dlomo, T. O., & Rogerson, C. M. (2020). Tourism and local economic development in King Sabata Dalindyebo local municipality, South Africa: Stakeholder Perspectives. *African Journal of Hospitality, Tourism, and Leisure, 10*, 145–164.

Donaldson, R. (2018). *Small town tourism in South Africa* (The urban book series). Springer.

Donaldson, R. (2021). Small town tourism in South Africa revisited. In J. Rogerson & C. Rogerson (Eds.), *Urban tourism in the global south* (pp. 193–211). Springer.

Donaldson, R., & Marais, L. (Eds.). (2012). *Small town geographies in Africa: Experiences from South Africa and elsewhere*. Nova Science Publishers New York.

Donaldson, R., Du Plessis, D., Van Niekerk, A., & Spocter, M. (2012a). Towards generic interventions to stimulate growth potential in small towns of the Western Cape Province, South Africa. *South African Geographical Journal, 94*(2), 120–136.

Donaldson, R., Spocter, M., Du Plessis, D., & Van Niekerk, A. (2012b). Towards generic interventions to stimulate growth potential in small towns of the Western Cape Province, South Africa. *South African Geographical Journal, 94*(2), 120–136.

Donaldson, R., Du Toit-Helmbold, M., & Bolton, A. (2021a). Towards repositioning the Swellendam municipal area as tourist destination: Tourism trends, marketing and branding. *African Journal of Hospitality, Tourism and Leisure, 10*(5), 1576–1590.

Donaldson, R., Eichhoff, J., Nightingale, F., Cloete, J., Ndamandama, M., & Erasmus, G. (2021b). Reflections on COVID-19 and its impact on marginal urban places: The case of four small towns in the Western cape, South Africa. In S. Brunn, D. Gilbreath, & D. Gilbreath (Eds.), *Covid-19 and an emerging world of ad hoc geographies* (pp. 1611–1630). Springer.

Drummond, F., & Snowball, J. (2019). Cultural clusters as a local economic development strategy in rural, small town areas: The Sarah Baartman District in South Africa. *Bulletin of Geography. Socio-economic Series, 43*, 107–119.

Drummond, J., Drummond, F., & Rogerson, C. M. (2021). Latent opportunities for heritage tourism in South Africa: Evidence from Mahikeng and surrounds. *African Journal of Hospitality, Tourism and Leisure, 10*(5), 1591–1609.

Dryer, L. (2013). *The role of urban and environmental parameters in sustainable development and future growth of small towns: Parys – A case study*. Masters dissertation. University of Free State.

Gibb, M., & Nel, E. (2007). Small town redevelopment: The benefits and costs of local economic development in Alicedale. *Urban Forum, 18*(2), 69–84.

Gwedla, N., & Shackleton, C. M. (2019). Perceptions and preferences for urban trees across multiple socio-economic contexts in the Eastern Cape, South Africa. *Landscape and Urban Planning, 189*, 225–234.

Hanekom, F. (1976). *Die hulpbronfront: 'n gevallestudie van delfstowwe se invloed op streekontwikkeling in Noorden Noordwes-Kaapland.* [Doctoral dissertation]. University of South Africa (UNISA).

Hebinck, P., Smith, L., & Aliber, M. (2023). Beyond technocracy: The role of the state in rural development in the Eastern Cape, South Africa. *Land Use Policy, 126*, 106527.

Hoefnagels, N., Irvine, P., & Memela, S. N. (2022). Makhanda: Exploring the mise-en-scène of a city under threat. *Urban Forum*, 1–21. https://doi.org/10.1007/s12132-022-09467-7

Hofstetter, M., Van Koppen, B., & Bolding, A. (2021). The emergence of collectively owned self-supply water supply systems in rural South Africa–what can we learn from the Tshakhuma case in Limpopo? *Water SA, 47*(2), 253–263.

Hoogendoorn, G. (2010). *Second homes and local economic impacts in the South African post-productivist countryside.* [Doctoral dissertation]. University of the Free State.

Hoogendoorn, G. (2021). Last chance tourism in South Africa: Future research potential? *Tourism: An International Interdisciplinary Journal, 69*(1), 73–82.

Hoogendoorn, G., & Fitchett, J. M. (2018). Perspectives on second homes, climate change and tourism in South Africa. *African Journal of Hospitality, Tourism and Leisure, 7*(2), 1–18.

Hoogendoorn, G., & Nel, E. (2019). Small town change in South Africa. In J. Knight & C. Rogerson (Eds.), *The geography of South Africa* (pp. 153–158). Springer.

Hoogendoorn, G., & Rogerson, C. M. (2015). Tourism geography in the global south: New south African perspectives. *South African Geographical Journal, 97*(2), 101–110.

Hoogendoorn, G., & Visser, G. (2010). The role of second homes in local economic development in five small South African towns. *Development Southern Africa, 27*(4), 547–562.

Hoogendoorn, G., & Visser, G. (2016). South Africa's small towns: A review on recent research. *Local Economy, 31*(1–2), 95–108.

Hoogendoorn, G., Mellett, R., & Visser, G. (2005). Second homes tourism in Africa: Reflections on the South African experience. *Urban Forum, 16*, 112–154.

Hoogendoorn, G., Grant, B., & Fitchett, J. M. (2016). Disjunct perceptions? Climate change threats in two-low lying South African coastal towns. *Bulletin of Geography. Socio-economic Series, 31*, 59–71.

Houghton, J., Dlamini, B., & Mthembu, N. (2013). Considering small towns research and local economic development in South Africa: Section I: Theory. *Skills at Work: Theory and Practice Journal, 6*(1), 13–25.

Jili, N. N., Masuku, M. M., & Nzimakwe, T. I. (2019). Enhancing pro-poor local economic development initiatives towards expediting sustainable development goals: King Cetshwayo District Municipality, KwaZulu-Natal. *Journal of Public Administration, 54*(4–1), 790–801.

Kativhu, S., Iwara, I. O., & Mwale, M. (2021). Exploring threats to youth enterprise resilience in selected areas in Thulamela local municipality, South Africa. *African Journal of Development Studies (formerly AFFRIKA Journal of Politics, Economics and Society), 1*, 119–144.

Khambule, I., & Gerwel-Proches, C. (2019). Exploring the role of social dialogue in local economic development: A case of a South African Local Economic Development Agency. *Progress in Development Studies, 19*(1), 36–54.

Khambule, I., & Mtapuri, O. (2018). Assessing the role of local economic development agencies in KwaZulu-Natal, South Africa. *Local Economy, 33*(4), 438–455.

Khapayi, M., & Celliers, P. R. (2016). Factors limiting and preventing emerging farmers to progress to commercial agricultural farming in the king William's town area of the eastern Cape Province, South Africa. *South African Journal of Agricultural Extension, 44*, 25–41.

Khumalo, S. M., Mthuli, S. A., & Singh, N. (2019). Economic development through the local informal economy in sustaining livelihoods: The case of the rural coastal town of Mtubatuba. *Journal of Public Administration, 54*(4–1), 772–789.

Kontsiwe, N., & Visser, G. (2019). Tourism as a vehicle for local economic development in small towns? When things go wrong: The case of Aliwal North, South Africa. *Geo Journal of Tourism and Geosites, 27*(4), 1334–1346.

Lawrence, F. (2013). The role of local economic development agencies in the south African local economic development landscape. *Urban Forum, 24*, 523–541.

Lawrence, F., & Rogerson, C. M. (2019). Local economic development agencies and peripheral small-town development: Evidence from Somerset East, South Africa. *Urbani izziv, 30*, 144–157.

Lombard, A., & Ferreira, S. L. A. (2015). The spatial distribution of renewable energy infrastructure in three particular provinces of South Africa. *Bulletin of Geography. Socio-economic Series, 30*, 71–85.

Mabin, A. (2021). History and hope in Cradock, Eastern Cape. *Transformation: Critical Perspectives on Southern Africa, 106*(1), 35–67.

Manomano, T., & Tanga, T. (2018). Housing needs: The quality and quantity of housing provided by the government for the poor in the Eastern Cape province in South Africa. *Social Work, 54*(1), 19–36.

Marais, L. (2013). Resource policy and mine closure in South Africa: The case of the Free State goldfields. *Resources Policy, 38*(3), 363–372.

Marais, L. (2023). *The social impacts of mine closure in South Africa: Housing policy and place attachment*. Routledge.

Marais, L., & Cloete, J. (2009). Mining and housing: The case of the "village under the trees" (Kathu – South Africa). *Town and Regional Planning, 55*, 32–38.

Marais, L., Campbell, M., Denoon-Stevens, S., & Van Rooyen, D. (Eds.). (2021). *Mining and community in the south African platinum belt: A decade after Marikana*. Nova Science Publishers.

Marais, L., Cloete, J., & Lenka, M. (2022). The plight of mining cities in South Africa: Planning for growth and closure. *Cities, 130*, 103965.

Marais, L., Ndaguba, M., Mmbadi, E., Cloete, J., & Lenka, M. (2022). Mine closure, social disruption and crime in South Africa. *The Geographical Journal, 188*(3), 383–400.

Masutha, M., & Rogerson, C. M. (2014). Small enterprise development in South Africa: The role of business incubators. *Bulletin of Geography. Socio-economic Series, 26*(26), 141–155.

McKibbin, J., Binns, T., & Nel, E. (2012). Uplifting small towns in post-apartheid South Africa: The experience of the Amathole Regional Economic Development Agency (Aspire). *Local Economy, 27*(4), 388–402.

Munien, S., Phungula, S., & Bob, U. (2018). Tourism potential, economic impacts and implications in marginalised areas in Ndwedwe Municipality, KwaZulu-Natal, South Africa. *EuroEconomica, 37*(2), 63.

Nel, E. (2005). Local economic development in South African small towns. In E. Nel & C. M. Rogerson (Eds.), *Local economic development in the development world: The experience of southern Africa* (pp. 253–265). Transaction Publishers.

Nel, S. (2017). *The development agency concept*. Paper presented at the National Local Economic Development Conference, Pretoria, 10 November. Typescript.

Nel, E., & Rogerson, C. M. (2007). Evolving local economic development policy and practice in South Africa with special reference to smaller urban centres. *Urban Forum, 18*(2), 1–11.

Ngcayi, Q. Q. (2013). *Service delivery, the state and NGOs – The case of Peddie extension, Eastern Cape*. [Masters thesis]. Nelson Mandela Metropolitan University.

Ntema, J. (2021). Alice: Socio-spatial change in a small university town. In A. Lemon, R. Donaldson, & G. Visser (Eds.), *South African urban change three decades after apartheid: Homes still apart?* (pp. 233–247). Springer Nature.

Oldfield, S., Anand, G., & Viviers, A. (2020). *Present and visible. Napier settlement and its stories*. University of Cape Town.

Roberts, S., & Shackleton, C. (2018). Temporal dynamics and motivations for urban community food gardens in medium-sized towns of the Eastern Cape, South Africa. *Land, 7*(4), 146.
Rogerson, C. M. (2012). Mining-dependent localities in South Africa: The state of partnerships small town local development. *Urban Forum, 23*(1), 107–132.
Rogerson, C. M. (2016a). Climate change, tourism and local economic development in South Africa. *Local Economy, 31*(1–2), 322–331.
Rogerson, C. M. (2016b). Outside the cities: Tourism pathways in South Africa's small towns and rural areas. *African Journal of Hospitality, Tourism and Leisure, 5*(3), 1–16.
Rogerson, C. M. (2016c). Outside the cities: Tourism pathways in South Africa's small towns and rural areas. *African Journal of Hospitality Tourism and Leisure, 5*(3), 1–16.
Rogerson, C. M. (2019a). Peripheral tourism trajectories: Evidence from the King Sabata Dalindyebo municipality, South Africa. *Geo Journal of Tourism and Geosites, 26*(3), 974–992.
Rogerson, J. M. (2019b). The evolution of accommodation services in a coastal resort town: Hermanus, South Africa. *African Journal of Hospitality, Tourism and Leisure, 8*(5), 1–16.
Rogerson, C. M., & Visser, G. (2014). A decade of progress in African urban tourism scholarship. *Urban Forum, 25*(4), 407–418.
Rogerson, C. M., & Rogerson, J. M. (2019a). Tourism, local economic development and inclusion: Evidence from Overstrand Local Municipality, South Africa. *Geo Journal of Tourism and Geosites, 25*(2), 293–308.
Rogerson, C. M., & Rogerson, J. M. (2019b). Tourism and accommodation services in South Africa: A spatial perspective. In *The geography of South Africa* (pp. 213–220). Springer.
Rogerson, C. M., & Rogerson, J. M. (2020a). Resort development and pathways in South Africa: Hermanus 1890–1994. In J. Rogerson & G. Visser (Eds.), *New directions in South African tourism geographies* (pp. 15–32). Springer.
Rogerson, C. M., & Rogerson, J. M. (2020b). Inclusive tourism and municipal assets: Evidence from Overstrand local municipality, South Africa. *Development Southern Africa, 37*(5), 840–854.
Rogerson, C. M., & Rogerson, J. M. (2022). The first round impacts of COVID-19 for rural tourism in South Africa. *Studia Periegetica, 1*(37), 63–86.
Rogerson, C. M., & Sixaba, Z. (2022). The limits of rural tourism in COVID-19 South Africa: Perceptions from 'left behind' rural spaces. *African Journal of Hospitality, Tourism and Leisure, 11*(1), 362–378.
Shackleton, C. M., & Blair, A. (2013). Perceptions and use of public green space is influenced by its relative abundance in two small towns in South Africa. *Landscape and Urban Planning, 113*, 104–112.
Shackleton, S., & Luckert, M. (2015). Changing livelihoods and landscapes in the rural Eastern Cape, South Africa: Past influences and future trajectories. *Land, 4*(4), 1060–1089.
Shackleton, C. M., & Njwaxu, A. (2021). Does the absence of community involvement underpin the demise of urban neighbourhood parks in the Eastern Cape, South Africa? *Landscape and Urban Planning, 207*, 104006.
Shackleton, C. M., Blair, A., De Lacy, P., Kaoma, H., Mugwagwa, N., Dalu, M. T., & Walton, W. (2018). How important is green infrastructure in small and medium-sized towns? Lessons from South Africa. *Landscape and Urban Planning, 180*, 273–281.
Simon, D., Vora, Y., Sharma, T., & Smit, W. (2021). Responding to climate change in small and intermediate cities: Comparative policy perspectives from India and South Africa. *Sustainability, 13*(4), 2382.
Social Work Policy Institute. (2006). *Housing*. [Online] Available: http://www.socialworkpolicy.org/research/housing.html. Accessed: 1/2/2018.
Spocter, M. (2016). Non-metropolitan gated retirement communities in the Western Cape. *Urban Forum, 27*(2), 211–228.
Spocter, M. (2017). Privatisation of municipal golf courses in small towns in the Western Cape, South Africa. *South African Geographical Journal, 99*(2), 113–133.

Toerien, D. F. (2018). The 'small town paradox and towns of the Eastern Cape Karoo, South Africa. *Journal of Arid Environments, 154*, 89–98.

Toerien, D. (2020). Tourism and poverty in rural South Africa: A revisit. South African. *Journal of Science, 116*(1–2), 1–8.

Toerien, D. F. (2022). Linking microeconomic characteristics, entrepreneurship and community prosperity/poverty of South African towns. *Development Southern Africa, 39*(2), 182–208.

Toerien, D. F., & Seaman, M. T. (2014). Enterprise richness as an important characteristic of South African towns. *South African Journal of Science, 110*(11–12), 01–09.

Tseane-Gumbi, L. A., & Ani, K. J. (2019). Tourism and transformation in the South Africa's former homelands: Lessons and implications for Africa. *African Renaissance, 16*(Special Issue 3), 93–108.

Tsheola, J. P., & Mokgokong, M. J. (2012). Public service delivery and local economic development in South Africa's impoverished communities. *Journal of Public Administration, 47*(1), 379–393.

Van Niekerk, A., Du Plessis, D., Boonzaaier, I., Spocter, M., Ferreira, S., Loots, L., & Donaldson, R. (2016). Development of a multi-criteria spatial planning support system for growth potential modelling in the Western cape, South Africa. *Land Use Policy, 50*, 179–193.

Van Rooyen, D. (2018). The effect of service delivery on local economic development in Brandfort, Free State. In A. Van der Walt, N. Kotze, C. Barker, & E. Kruger (Eds.), *Proceedings of the Society of South African Geographers* (Vol. 1, pp. 103–117).

Visser, G., & Hoogendoorn, G. (2011). Current paths in south African tourism research. *Tourism Review International, 15*, 5–20.

Visser, G., & Marais, M. (2021). The impact of COVID-19 on safari lodges in Northern KwaZulu-Natal, South Africa. *African Journal of Hospitality, Tourism and Leisure, 10*(5), 1683–1694.

Welman, L., & Ferreira, S. L. A. (2016). Development of the Greater Saldanha Bay Region: The co-evolution of Saldanha town and its harbour. *Local Economy, 31*(1–2), 219–234.

Zama, S. C. (2020). *Assessing the impact of low-cost housing in small towns: a comparison approach of the Nkandla and Umzimkhulu Municipalities in South Africa* [Doctoral dissertation]. University of Kwazulu-Natal.

Zhang, H., Xu, Y., & Kanyerere, T. (2019). Site assessment for MAR through GIS and modelling in West Coast, South Africa. *Water, 11*(8), 1646.

Zulu, K. C., & Mubangizi, B. C. (2014). Rural local economic development: Insights from a South African rural community. *Journal of Public Administration, 49*(1), 424–438.

Chapter 2
The Relationship Between White Privilege, Segregation, Topophobia and Symbolic Boundaries in Secunda, Mpumalanga

Tarryn N. K. Paquet

2.1 Introduction: Introducing the Link Between Symbolic Violence, Topophobia and White Privilege

Despite more than 20 years of democracy, South African planners and policymakers continue to face the many challenges, complexities and 'stubborn realities' of transforming grossly unequal and segregated urban forms (Visser, 2001; Harrison et al., 2008; Newton & Schuermans, 2013; Watson, 2013). I argue that these challenges and complexities do not end with physical inequalities and injustices, but also include abstract realities such as symbolic violence and the subjectivities that lie at the root of symbolic violence. For the purposes of this chapter, symbolic violence occurs not when injury is caused to a person or property, but through the control of emotional well-being.

To a large extent, symbolic violence has become an almost taken for granted part of everyday South Africa. Lefebvre (1971) argues that, when symbolic violence becomes taken for granted, it is no longer recognised by victims and oppressors alike, and its effects lead to a 'terrible unease'. This, in turn, can bring about a latent form of fear, which resides in the subconscious (Lefebvre, 1971). For the purposes of this study, the use of 'latent' will be used in the sense that something remains

---

Current affiliation of employment
This chapter is part of a PhD thesis research conducted under the University of Cape Town, 2012 to 2017

T. N. K. Paquet (✉)
Department of Architecture and Urban Planning, Qatar University, Doha, Qatar

School of Architecture, Planning and Geomatics, University of Cape Town, Cape Town, South Africa
e-mail: t.paquet@qu.edu.qa

© The Author(s), under exclusive license to Springer Nature Switzerland AG 2023

R. Donaldson (ed.), *Socio-Spatial Small Town Dynamics in South Africa*, GeoJournal Library, https://doi.org/10.1007/978-3-031-37142-4_2

hidden and dormant until suitable situations arise for it to manifest ("Latent, adj.", 2014). I consider this latent form of fear in its space-based capacity – known as 'topophobia'. The term 'topophobia' was first coined by Tuan (1974, 1979) who saw space as being subjective and suffused with emotion. He considered – through the binary concepts of topophilia (love of place) and topophobia (spatial fear) – the meaning of space on the attitudes and emotions of people, and how they, in turn, influenced space.

Topophobia has a particular significance for urban planners as the creators, administrators, and users of urban space. Fear affects us all and greatly influences the manner in which we consider and shape urban space (Sandercock, 2005; Meth, 2016). Thus, all spaces are 'vulnerable' to topophobia. Arguably, if we are all susceptible to topophobia, the space we create must also be vulnerable to the condition. As such, ignoring emotions – such as fear – that affect us as planners, as well as individual users of the spaces we create, will be to ignore many of the issues we are faced with as a country.

Lefebvre (1971) makes a strong link between 'unease' (or anxiety), fear and symbolic violence. Briefly, he indicates that a privileged group might rely on symbolic violence to maintain 'systems' which favour the powerful. Arguably, one such 'system' in South Africa is white privilege. White privilege is a system of dominance geared towards preserving an advantageous lifestyle for a powerful group, usually to the detriment (whether consciously exploited or not) of 'other' groups (Steyn & Foster, 2008). The symbolic violence, and resultant topophobia, of such systems are so subtle and consistent that the fear they produce becomes ingrained in the subconscious, making it latent (hidden or dormant). My research seeks to uncover and come to grips with this problem, which is 'wicked' – but not in the sense that it cannot be unravelled and confronted.

This chapter seeks to address the issue of white privilege and its relationship to topophobia, by exploring how Secunda, a former apartheid new town, has transformed into 'comfort enclaves' for white people in an effort to preserve their privileged status under the current democratic government. The chapter's objective is to examine the ways in which these enclaves are upheld by symbolic boundaries, such as market filters and distancing attitudes.

To do so, it commences with a review of the existing literature on the relationship between white privilege, topophobia and symbolic violence. Lefebvre's (1971) theory on the symbolic violence of what he calls "terrorist societies" is used as a theoretical underpinning. As a former outpost of the independence of the apartheid government in the wake of international oil sanctions, the situated context of the new town of Secunda provides a rich opportunity for understanding topophobia and symbolic violence (both in South Africa and elsewhere). The chapter then turns to the research methods employed, followed by a demonstration showing that Secunda remains an enclave of (white) privilege, which Schensul and Heller (2011) indicate, are becoming more frequent across South Africa's urban landscape due to an increasingly class-based society.

Further findings show how topophobia, and resultant symbolic violence, evolve in the loss of 'comfort zones' within enclaves of whiteness. Findings also indicate that dissociative attempts may be coupled with symbolic boundaries designed to further distance, or divide, 'the same' from 'the other'. Distancing can be accomplished through distinctions made in the everyday between the "threatened Self" and the "threatening Other", which are "fundamental in the white struggle for Selfhood" (Hubbard, 2005: 60). The next section will commence with a review of the literature on the relationship between 'othering', white privilege and symbolic violence.

2.2 A Review on the Relationship Between White Privilege, Topophobia and Symbolic Violence

For the purposes of this chapter, 'othering' refers to the labelling of persons who cannot, or will not, conform to specific social codes that favour a certain group (which is usually privileged and powerful). 'Othering' and topophobia are inextricably linked, says Koskela (2010), as without the other, there would be nothing to fear, and without fear, there would be no the other.

Further, topophobia breeds the inability to accept difference (Janz, 2008), ruptures social relationships, decreases quality of life, and blocks both social and physical integration as "[w]ithout trust, there is no interaction" (Koskela, 2010: 403). An important point must be made here. According to Lefebvre (1991), social relationships make up the perceived and lived space of cities. Thus, it can be argued that it is here that our lived experience emerges, influencing the way in which we produce place identity and infuse our urban space with meaning. If social relationships are ruptured by a fear of 'the other', urban transformation will be very difficult to effect no matter how well we, as planners, attempt to conceive of topophilic[1] cities that embody equality. Therefore, a greater understanding of the topophobias prevalent in social relationships, as well as the reactions they produce, is needed to reach the goal of transformation in South African cities.

Othering is usually based on characteristics of identity such as poverty (Sandercock, 2005; Whitzman, 2007), gender (Whitzman, 2007), and ethnicity or race (Koskela, 2010; Yiftachel & Yacobi, 2003). Markedly, fear of 'the racial other' has a large impact on contemporary South African cities as a result of apartheid planning and continues to "dominate sociospatial exclusions''within the urban realm (Lemanski, 2004: 103).

One of the main drivers behind this can be attributed to the patriarchal ideology of apartheid, which developed hierarchies along race, gender and ethnic divisions (Freund, 2013; Sparks, 2012). This ideology ultimately sought to instil a consciousness of whiteness, which sought (and seeks) to ensure the progression of white

[1]This dichotomy stems from Tuan's (1974, 1979) original definitions of topophobia/topophilia during a time when modernist planning was at the height of its popularity and the pitfalls of seeing the world in terms of binaries had not yet surfaced.

power and domination (Ballard, 2004; Steyn, 2004). It is important to note that, in keeping with Friedmann's (2010) principle of interiority (where multiple identities are produced by the subjectivities of individual space-users over time), multiple identities of whiteness can inhabit a single space. This argument is supported by Hubbard (2005) who continues that some identities of whiteness may not all be subject to equal forms of white privilege (for example, the late nineteenth-century Roma and Traveller people in the United Kingdom who were/are exposed to social discrimination).

Although not the first to initiate segregationist policies, the apartheid government created what Ballard (2004: 51) calls 'comfort zones' to maintain a Western, modernist place identity to bolster whiteness within an African context:

> *[T]he minority government removed those people, values, behaviours, languages which were seen to contradict this identity.*

Ultimately, argues Ballard (2004: 54), white people used exclusion and segregation to "contend that they lived in civilised, modern, First World cities." This view can be related to a modernist quest for order and social reform, which saw the displacement of 'the other' as a necessary price in the overall public interest: "[t]here was really no good reason to tolerate the Other who, by definition, rebelled against the truth" (Bauman, 1992: xiv). To further ensure segregation between spatial zones, while simultaneously cementing a dominance over spaces of privilege, a dichotomy of 'the white same' and 'the black other' was established[2] (Ballard, 2004).

This dichotomy sought to produce 'superior' or 'civilised' identities by equating white people with labels such as 'order', 'purity', 'modern', 'clean', 'moral' and so forth (Ballard, 2004; Hubbard, 2005). In keeping with this thinking, black people were depicted as "lazy, licentious, criminal, dirty, and so on" (Ballard, 2004: 52). These fabricated identities have proven to be powerful tools in the maintenance of whiteness and white privilege in post-apartheid South Africa and elsewhere – see "'Losing' Comfort Enclaves" below for an example in Secunda.

Arguably, symbolic violence is more likely to be accompanied by symbolic boundaries in post-apartheid contexts, where boundaries separating races are no longer state-condoned. In fact, Spinks (2001) goes as far as to contribute symbolic, rather than physical, boundaries as apartheid's most lasting legacy. She argues that state propaganda, which encouraged race groups to think of themselves as 'separate', remains entrenched in many post-apartheid contexts. A participant, 'Mienkie's' (translated 18/03/2013) comment asserts that understandings of 'separateness' are often entrenched in various whiteness identities from a young age:

> *That's just how white people, especially Afrikaners, are raised: That you don't mix. That's just how it is.*

This 'schooling' is one reason that racially-based topophobias and biases are often latent, or deeply embedded in the subconscious. Further, with the location of

[2] Notably, this dichotomy is not pertinent only to South Africa, but may affect most Western cultures and identities of whiteness (Hubbard, 2005).

whiteness and its biases lying in the subconscious of white people, the latent and entrenched codes of white privilege often manifest as 'knowledge', 'truth', or value systems (based on statements of perceived 'fact' such as 'that is just how it is'). This, in turn, can lead to an ignorance of the racially-based privileges which many white people benefit from. As such, many white people fail to recognise when they are being racially offensive (Hubbard, 2005), and may not recognise the symbolic violence they could be meting out in the assertion of whiteness and white privilege. Further, opposition to transformation efforts may find its roots in identities of "unmarked whiteness" and the maintenance of white privilege (Hubbard, 2005: 52).

The study of white privilege follows a path that is both viscous and messy at the same time. A methodological framework that is reflexive and can tackle the complexities of systems of dominance is needed. The next section sets out the methodology for and lessons learned from the path walked during this research.

2.3 Research Methods

This chapter draws from a larger study aimed at understanding symbolic violence and spatial fear (topophobia) within a situated urban context. Due to the emphasis on emotion-based research, a feminist ethnography based on qualitative data was deemed most suited to uncovering the complex power struggles that influence the topophobias of women in Secunda. A feminist ethnography can be seen as a form of case study (which is a popular method in planning research) as both seek to provide a 'thick' description and challenge past assumptions (Flyvbjerg, 2001). However, I have chosen to concentrate on ethnography as this method has proven to be a powerful tool in teasing out the impact of gender identities on space (Greed, 1994).

A feminist ethnography requires that trust and relationships be built with participants (Stacey, 1991). Relationships of trust may serve as platforms to uncover and explore the deep-seated emotions tied to urban space. If these platforms are made use of, it is recommended that this is done honestly, with great sensitivity and within everyday contexts as they occur. Thomas (2003) cautions that, in attempting to build relationships of trust with participants, the researcher can become too immersed in the field, thereby decreasing the rigour of the study through bias. I found that repetitive interview techniques are especially vulnerable to this weakness. As such, when using a feminist ethnography, it is recommended that multiple techniques be employed to allow the researcher to desist from further interviews in cases of over-emersion.

As such, both structured (4) and conversation-type (25) interviews with 20 participants (including those involved with the initial construction of the town, municipal officials, and residents) were undertaken in addition to participant observation, observations of Secunda's digital space (specifically, Facebook, blogs and news articles), books written by residents, and archival data. Two archives were drawn upon, namely, the Historical Papers Archive located at the University of Witwatersrand, and the archives of Sasol's Real Estate department in Secunda.

Notably, most of the newspaper articles referenced were collected from these archives. Follow-up conversations were also held with participants via email and text message, and informal conversations in the streets, shops and homes of Secunda's residents were recorded in my research journal as part of my observations. An iterative field research process was employed over a period of two and a half years. Here, (*inter alia*) transcription took place in conjunction with interviewing, allowing for the identification of follow-up interviews where necessary.

Findings were analysed and interpreted through an inductive, themed-based analysis using Atlas.ti software. Discourse analysis was also drawn upon as a tool to determine the way in which 'truth' is constructed from the meaning of Secunda's space. All names referred to in the findings are pseudonyms and have been changed to protect the identities of participants. In the instances where interviews or discourse analyses were originally conducted in another language but translated, these have been highlighted in a footnote.

My own multiple insider identities (of middle-class, white, English-Afrikaans woman, and planner) were used to critique systems of dominance, including white privilege, and to gain an in-depth understanding of the symbolic boundaries some may erect in a bid to maintain power status quos. These multiple insider identities allowed greater insight into the nature and meaning of Secunda. However, it is important to note that, while multiple insider identities are a strength, they are also a limitation of this research. In other words, this research seeks to contribute to global South literature by providing a deeper understanding of white privilege from a white person's perspective. As Sihlongonyane (2015) has made apparent, it does so through the use of Western epistemology, theory, and methodology. I by no means dispute the value of understanding systems of dominance, multiple and intersecting identities such as the victim-oppressor within these systems as well as the contribution that such an understanding may have on unravelling certain oppressive systems within South African space. In fact, this understanding has cemented a conviction that white privilege is a pervasive and serious urban planning problem.

But, in focussing so strongly on my multiple-insider identities within the white communities of South African space, I inadvertently blinkered myself. Specifically, the missing voices of black South Africans in this study leaves it with a sense of indebtedness, and I name it as a limitation of this study. This limitation was fuelled by an anxiety of my own brought about when I realised my white ignorance and the resulting quest to unravel the concept of white privilege. I never set out to explore the link between topophobia and white privilege. In fact, I tried to avoid it on many occasions. But that is the strength of an inductive study – it lets the context dictate what should be studied. The relationship between topophobia and white privilege was simply unavoidable. Upon coming to terms with this, my initial standpoint was that white privilege is a white people's problem and had to be solved within this group. However, I would like to caution future researchers against taking this stance as it negates many nuances and falls into the trap of dichotomous thinking. White privilege affects many different societies and should rather be tackled from a unified front. I now turn to the situated context of Secunda, which proved so rich in its complexity.

2.4 Secunda: An Enclave of Whiteness and Segregation

Secunda, located in the Mpumalanga province (Fig. 2.1), is the second (the first being Sasolburg, Free State) company town that belongs to Sasol Ltd[3] – an industrial company that specialises in the Fischer-Tropsch coal-to-liquid process (Meintjies, 1975; Sparks, 2012) unique to South Africa (John, 2012).

As a grassroots development, it is also a new town built in response to international oil sanctions in the 1970s under the apartheid regime (Freund, 2013). Residential segregation is a serious, and continuing, urban issue across South Africa (Durrheim & Dixon, 2010; Schensul & Heller, 2011). Secunda is a particularly glaring example made apparent from a comparison between the town and its larger (national and regional) environment (Fig. 2.2).

According to Fig. 2.2, white people are in the minority in larger contexts (16% of the municipal area, 7% in the Mpumalanga Province and 9% in South Africa) – a vast contrast to the dominance prevalent to Secunda at 72% of the town's

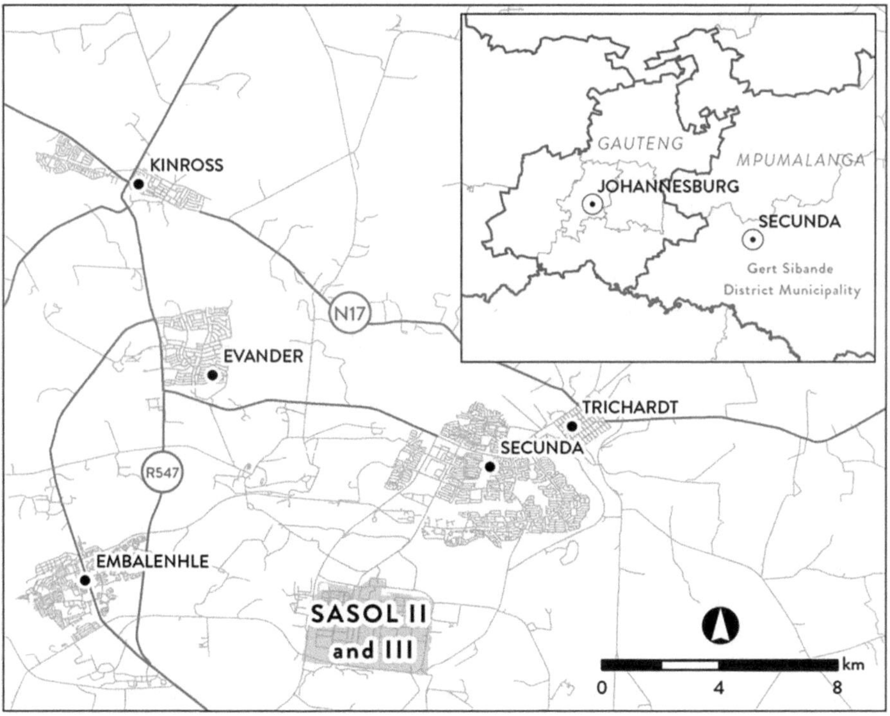

Fig. 2.1 Secunda and its region

[3] Its common name: The acronym stands for South African Synthetic Oil Limited.

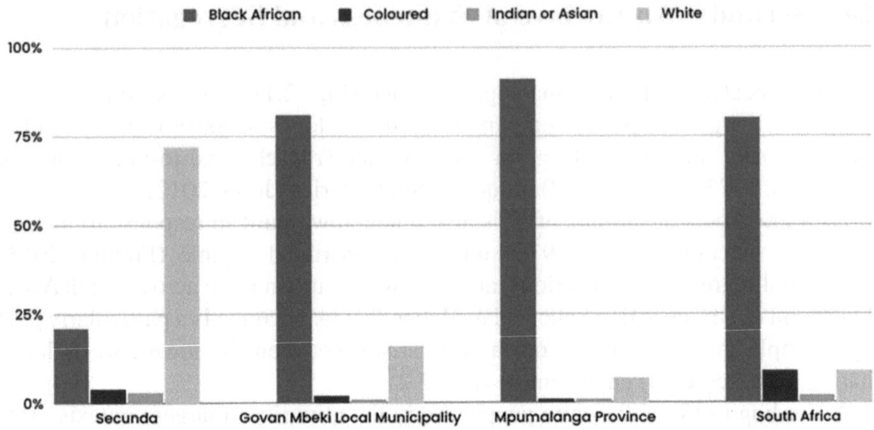

Fig. 2.2 A comparison of population percentages. (Source: Adapted from StatsSA (2011))

Table 2.1 Population group by size in Secunda[a]

Year	Black African		Coloured		Indian or Asian		White		Total
	No.	≈%	No.	≈%	No.	≈%	No.	≈%	
2001	6218	23.8	424	1.6	395	1.5	19,088	73.1	26,125
2011	8283	20.7	1508	3.8	1225	3.1	28,910	72.4	39,926
Growth[a]	2065	33.2	1084	255.7	830	210.1	9822	51.5	52.8%

Source: Adapted from StatsSA (Census 2001, 2011)
[a]Percentages rounded off to the first decimal

population.[4] Further, Table 2.1 indicates that Secunda has grown by over half its population size between 2001 and 2011, with the greatest growth *percentage* experienced amongst coloured and Indian/Asian population groups.

However, despite this, segregation levels have barely shifted over this period where the dominant racial group's presence has shifted only slightly from 73.1% (in 2001) to 72.4% (in 2011) of the overall population.

These figures denote a dual city phenomenon (as a remnant of apartheid planning) where the formerly 'white' town of Secunda remains segregated (both socially and physically) from its former 'black' 'township'/'location', eMbalenhle, whose population is dominated by black people at roughly 99% (StatsSA, 2011).

This physical segregation can be traced to the adaptation of the Burgess concentric zone model to reflect the spatial ideologies of apartheid (see Davies, 1981 for a full description of this). Referred to as the apartheid city, this adaption embraced the theory that "race and cultural differences in society are incompatible and that contact between ethnic groups leads to friction" (Davies, 1981: 69). What emerged from this involved individual race areas segregated by buffer zones or boundaries,

[4] It is noteworthy to add that the latest census (2022) was being conducted at the time of publication of this chapter and that segregation trends in South Africa would benefit from further research and analysis of these figures.

and controlled movement lines (usually for public transport) that, where possible, provided direct links between work centres and former 'black townships' (Davies, 1981).

Apartheid planning achieved its infamy through these segregated 'locations' or former 'black townships' to which black people were forcibly removed (Mabin & Smit, 1997). These segregated townships developed when valuable land was labelled as a white domain while black people were relegated to impoverished and overcrowded 'locations' on the outskirts of the city (Christopher, 1983; Sihlongonyane, 2002). Large housing estates, located far from white residential areas were built (Christopher, 1983; Mabin & Smit, 1997) to decrease contact between races (Christopher, 1983) while also establishing a labour supply to nearby urban areas (Mabin & Smit, 1997; Oranje, 1998). Races were further prevented from integrating by means of buffer zones (Christopher, 1983; Sihlongonyane, 2002) which took the form of roads, green spaces, factory sites, railroads, and even fencing.

Evidence of the apartheid city model is apparent in Secunda. Figure 2.3, an original rendition of the town's spatial structure (which remains largely unaltered today), shows the former 'white' area of Secunda, separated by a large buffer zone (approximately seventeen kilometres by fastest route) from eMbalenhle, the former 'black township' related to it.

As one of the most prominent symbols of modernity in apartheid South Africa, Max Kirchhofer, the town's designer, attributes its influences to the models of garden cities, British new towns, and the model of Radburn, USA (Kirchhofer, 1982). However, Secunda maintains strong influences of company town planning too. For example, in traditional company towns, social hierarchies went hand-in-hand with paternalistic ideals and were imposed through housing policies (Baxter, 2012). Typically, larger and prettier houses (placed in the public eye) were allocated to middle-class employees while working-class employees were located on the outskirts of the town in smaller housing units (Baxter, 2012; Rees, 2012). In the case of Secunda, high-tier Sasol employees were allocated houses in what is fondly referred to as the "Green Strip" or "Green Area" among residents (journal notes). The most prominent of these employees were afforded houses adjacent to the Duck Pond, a pondscape located in the 'Green Area' (Fig. 2.4).

Kirchhofer shared many new town planners' captivation with duck ponds (Greed, 1994). Secunda's 'Duck Pond' was designed to be the most 'superior' (or highest tier) of all the green strips. It comprises a set of small dams located in the southern valley of the town, which offer non-motorised water sports, fishing, scenic walkways, a mini train track, and picnic facilities to promote outdoor living (Wessels, 1990). In keeping with paternalism, company town, modernist and apartheid hierarchical ideologies, Secunda's green network was also used to demarcate the status of classes:

> *[Y]our more wealthy people lived in the lower lying areas closer to the public open spaces. So if you had your house it would look out onto the dam or out onto a park.*
> *(Bert, 15/04/2015)*

Bert, who was involved in the grassroots development of the town, indicates that Sasol's elite was housed around the Duck Pond and in its adjacent neighbourhood.

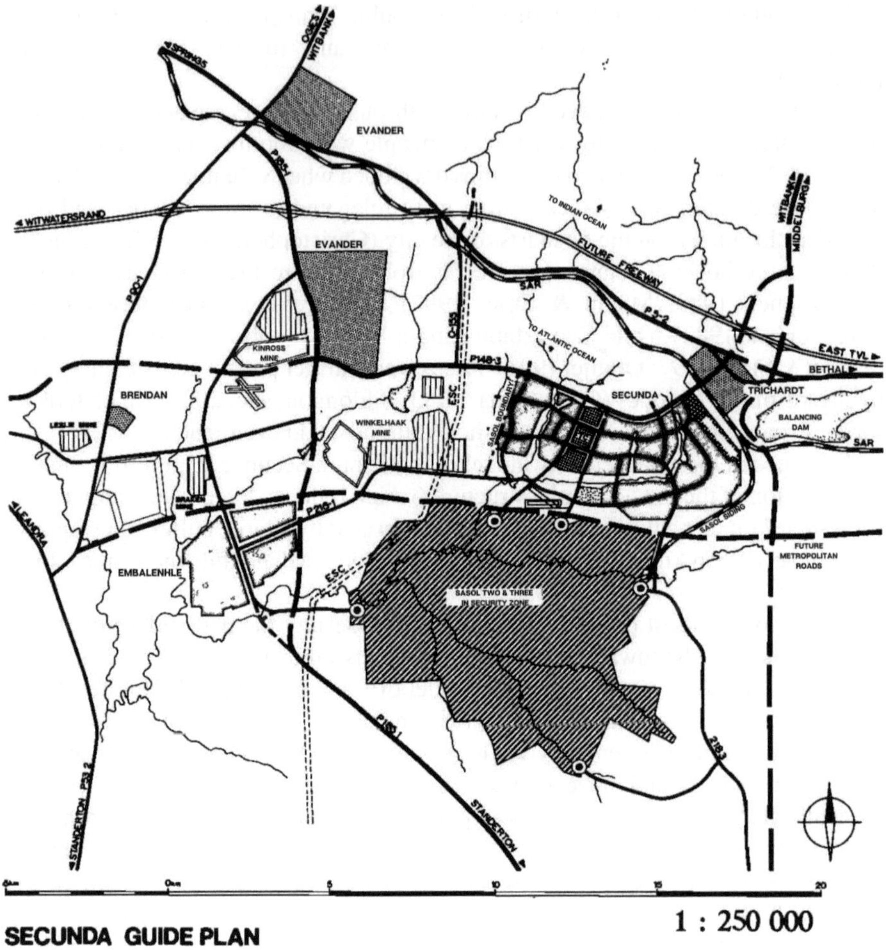

Fig. 2.3 An original plan of Secunda's regional layout. (Source: Adapted from Kirchhofer (1982: 22))

Although Sasol no longer allocates housing for purchase, these houses have maintained high values, remaining accessible only to high 'level' employees and wealthier residents. The next section will reflect on the interiority of the Duck Pond, along with its commercial counterpart – the CBD, as past pillars of the former white town of Secunda.

2.5 Losing Comfort Zones

Through its high levels of segregation (where white people constitute 72.4% of the total population as shown above), Secunda is arguably an enclave of persistent whiteness. Although a prominent example, this phenomenon of creating or

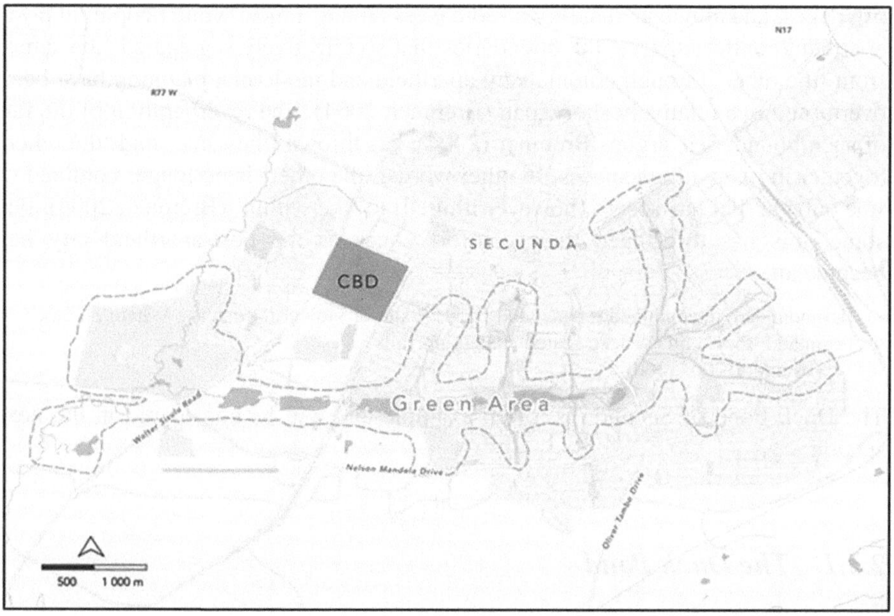

Fig. 2.4 The location of Secunda's CBD and green area

maintaining white comfort enclaves is not particular to Secunda (Ballard, 2004; Schensul & Heller, 2011). Ballard (2004) considers the contemporary burgeoning of these enclaves, which he says, has become largely privatised. He refers to this as 'semigration' that originally described the migration of white people to Johannesburg or Cape Town. However, he continues, semigration can also occur when high walls are erected around a property, or through the migration to gated communities or estates. Semigration is an attempt to re-establish "a comfort zone that reflects" the "self-conceptions"of some white people (Ballard, 2004: 65). I would argue that instead of 'semigrating' to fortified enclaves or other provinces, many of Secunda's residents seek to maintain Secunda as a predominantly white comfort enclave by erecting and/or maintaining various symbolic boundaries (which are maintained through instances of symbolic violence).

For example, Teresa (11/02/2013) posits:

> *It's a very segregated society. Hugely so. Like I think this is the closest to getting a snapshot of being like in the 1970s with the animosity or the superiority that white people still feel towards black people. You know you seldom see an old black person here that is on the same [salary] level and I don't know if that's got anything to do with violence but it's a very strong white identity here. Um ja, I guess in a way, although not maybe like physical violence, more like violent feelings towards like people.*

Teresa's comment hints at a relationship between deep-seated fears of 'the racial other' and an attempt, by some identities of whiteness, to maintain a 'comfort zone' by means of symbolic violence. Ballard (2004) considers the concept of white 'comfort zones' established through certain social, economic and psychological privileges, and aimed at ensuring a rootedness amongst most white people. In post-apartheid comfort spaces, the boundaries that so effectively segregated 'the same' from 'the other' through colonialism, apartheid and modernist planning have been overturned in a relatively short span (Bremner, 2004). The reverberation of the fall of such boundaries, argues Bremner (2004), has thrown 'the same' and 'the other' together in many urban spaces. In other words, 'the other' is no longer confined to 'the outside' but has access (however limited) to 'the within' (Bremner, 2004). For some, this has threatened their comfort zones as the post-apartheid city has become an:

> [U]nbounded, uncontained, openended body. A site of violent intimacies. Where do 'we' begin and 'they' end? A deep seated anxiety prevails.
> (Bremner, 2004: 460)

The Duck Pond of Secunda is a rich example and will be considered in the next section.

2.5.1 The Duck Pond

> *I don't like the Duck Pond at all since it became so black and degraded. People tell me that it is nice to walk along the paths now and that there are braai [BBQ] facilities. Years ago, we would go there on Sunday mornings and throw bread to the ducks. It was lovely then, but we've never gone back since it became so scary there. My husband is very strict...he is...I won't call myself a racist – I always say, don't do anything to me and I won't bother you – but we [black and white people] didn't grow up in this town together and it is just not nice for us. We don't actually want to be where they are, we don't enjoy it.*
> *(Loraine, translated 26/03/2015)*

Although Loraine acknowledges recent attempts to remove signs of decay in the Duck Pond, her topophobia stemming from its forfeiture as a formerly 'white' space. Under the previous dispensation, only white people were allowed to frequent the Duck Pond. With the repeal of the Group Areas Act, former 'white' areas are no longer restricted, and the Duck Pond is now home to a diversity of visitors who eagerly make use of its facilities. This disruption of white dominance has resulted in a drastic shift in power (struggles for space) and the presence of the racial other has shifted the original social order (Sandercock, 2005). Durrheim and Dixon (2010) explain that some white people may experience the desegregation of previous 'white' spaces as a loss of their comfort enclaves. In this case, Loraine's unrequited urge to expel 'the other' from her comfort enclave has resulted in her abandoning the Duck Pond. Other examples of racially-based topophobia are evident in Secunda's green strips:

> *They don't use toilets; the black women simply lift up or pull down their clothes and do their business right there in the green strip.*
> *(Suzette, translated 15/01/2015)*

Suzette's indignation is not uncommon or new. Sparks (2012: 161; 164) records similar complaints made in Sasolburg (in the 1960s and 1970s):

> *I think it's objectionable. Every day, in your parks and gardens you can see Bantu men and women under the trees and on the grass, lying in the shade of the trees that you have planted for beautification.*

> *Every time I come out of my front gate I see a native urinating against a tree. It's not something that one wants to speak of, but how can one go and kick up a fuss with such a native? I am embarrassed to speak with you about this, even just on paper, but things cannot go on like this.*

Notably, this may point to a lack of public amenities in Secunda's green areas. For example, feeder stops (both formal and informal) for local minibus taxis are generally located near green strips, which were designed by Kirchhofer (1982) as a network to facilitate pedestrianism. Although some of the green spaces do provide public toilets, these are often associated with a lack of maintenance and/or crime. This is a debilitating situation that is beyond the scope of this chapter but requires further research and design interventions. Another example where a perceived loss of comfort enclave has occurred is the town's central business district, or CBD. This will be considered next.

2.5.2 *The CBD*

The layout of the town centre of Secunda forms a perfect square (Fig. 2.5). It consists of the central business district (CBD), and a civic centre including municipal buildings, a post office, magistrate's court, theatre and library. It is noteworthy that the municipal grounds have more recently expanded to include the old hotel.

Once a centrepiece of white dominance, the town centre is perhaps the most prevalent example of racially-based topophobia, for white people, in post-apartheid Secunda. Designed with a strict nod to the modernist planning ideal of order and formality, the CBD has organically evolved into a mix of both formal and informal activities. One of the latter involves spaza caravans. These spaza, or informal shops, are usually of a micro-economy and cater mainly to working class black people in Secunda. Figures 2.6 and 2.7 are artists' rendition of spaza shops operating out of caravans located on the outskirts of the Secunda CBD.

These spaza caravans are similar to the food trucks found worldwide. These food trucks are a growing trend in other parts of the world where they are linked to words such as 'haute' and 'cool culinary trend'. Although their informality is also seen as an urban issue by some, this is largely attributed to the competition they provide to formalised, tax-paying businesses (see for example, Gustin, 2010). However, this is not always the case in Secunda's CBD. In the following excerpt, a female resident expresses her unease at the increase of spaza shops housed in caravans in the CBD.

> *Hi Bianca, I spoke to you last year about the caravans stationed in front of Telkom/the Gym in town [the town centre]. When are those things [the caravans] going to be removed? I'm*

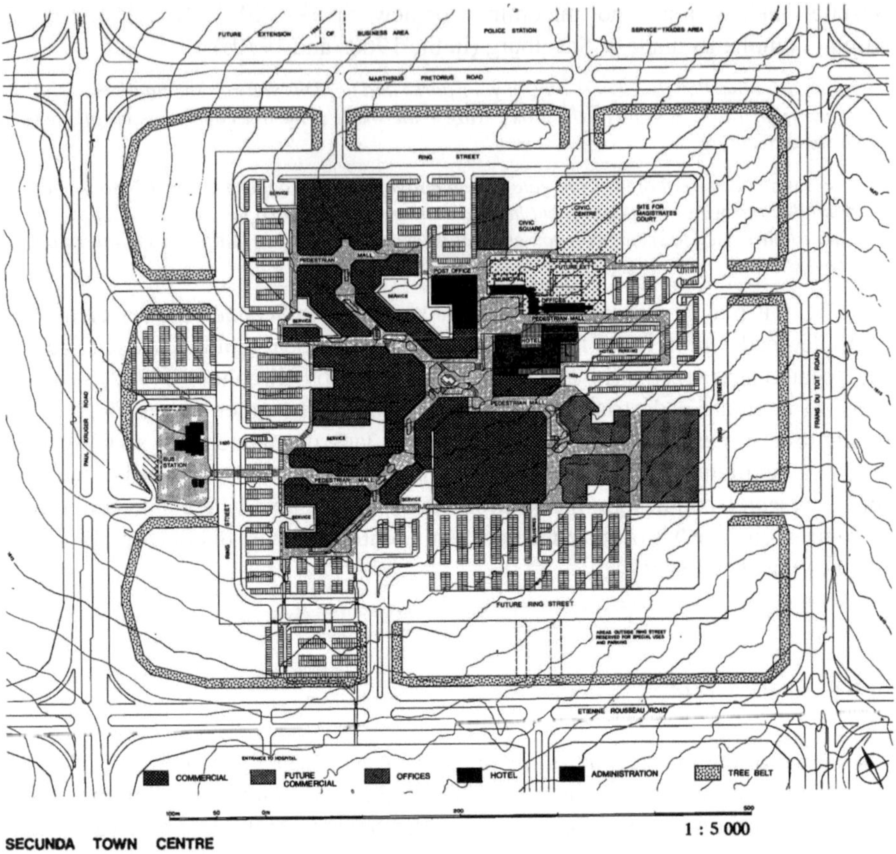

Fig. 2.5 An original plan of Secunda's town centre. (Source: Kirchhofer (1982: 25))

> *not even prepared to invite my friends from overseas to Secunda: we have to meet them in the [nearest] city! It looks like we stay in North Africa. They are unhygienic and it looks like we do business in a squatter camp. Must we, as Afriforum [a civil rights organisation, which focusses largely on Afrikaans culture], make our presence known or what do you suggest as councillor of the business district? I am out of ideas! Thank you!*
>
> *Trienkie, it is really not an easy situation to solve. It actually appears as if they are escalating in number. I have already opened a case, but it doesn't seem as if we are going to get a solution anytime soon. We could perhaps approach this matter from different sides. I would like to discuss this matter with you in more detail. Can we meet next week?*
> *(Facebook update, January 2013)*

Trienkie's topophobia at the increased presence of informal spaza shops has led to an attempt to mobilise other like-minded organisations (like Afriforum to which she belongs and the ward councillor, Bianca). Phrases such as 'make our presence known' and 'approach this matter from different sides' are evidence of symbolic violence (to be) inflicted in an attempt to reassert dominance of the town centre. The driving force behind modernist planning was a rationalisation of space including

Fig. 2.6 A sketch of the spaza caravans located in the Secunda CBD, by Noor Elbadri (2023)

Fig. 2.7 Making Amagwinya (or 'vetkoek' which is a yeasty bread fried until crisp and sometimes served with a filling), by: Zahra Al-Najar (2023)

determined efforts to eradicate informal spaces (sometimes referred to as 'disorderly' spaces) (Baeten, 2002; Sandercock, 2005). This rationalisation was infused with (male) Western, middle-class ideologies that viewed conflicting approaches/individuals as 'irrational' and 'undesirable' (Sandercock, 2005). Further, attempts to displace 'the other', as a means of maintaining 'order', were often based upon justifications of 'public interest' (Friedmann, 1987).

The 'breakdown' of modernist order in Secunda's CBD (for example, where 'categorised zoning' is permeated with 'informal' activities) has aggravated topophobia for some of its residents. In the excerpt above, the public interest is seen to be 'jeopardised' through allegations of unsanitary conditions. Sparks (2012: 161), who has based his research on the first Sasolonian town of Sasolburg, argues that the presence of 'the other' in comfort enclaves is often "linked to 'matter out of place' – to dirt, to litter". This assertion is arguably a form of symbolic violence with the intent of reinforcing symbolic barriers between the racial other ('It looks like we stay in North Africa... it looks like we do business in a squatter camp') and 'the white/Western same'. Concern at what appears to be diminishing the dominance of white people over the town centre is also expressed ('it actually appears if they are escalating in number').

The breakdown of racial divides and desegregation in Secunda's town centre, coupled with failed attempts at white dominance in this area have produced topophobias: most of which are centred on 'the racial other':

Tarryn: Do you feel safe in the town centre?
Karla: I never go to the Pick n Pay Mall; I'll rather go to the Checkers Mall and that area near Mr Price.
Suna: It feels more protected there.
Karla: I'm not the type of person who goes to Pick n Pay Mall. I hardly ever go there.
Tarryn: Why not?
Karla: Ohhh, it was never an attraction for me.
Suna: And you're a bit out of place there.
(Suna and Karla, translated 25/02/2013)

The design of Secunda's CBD encompasses various extensions, colloquially known as 'malls' The 'Checkers Mall' referred to is located adjacent to Sasol's gym and had, before the regional mall opened in November 2013, a larger white patronage than the then 'Pick n Pay Mall', which is closer to the taxi rank. Notably, the Secunda taxi rank caters largely to the black working class. Conversely, membership to the recently modernised/upgraded Sasol gym is inexpensive as a benefit for employees and their families – who were traditionally encouraged to be 'respectable citizens'. Although this is a topic beyond the scope of this chapter, it can be summarised that desired identities for Secunda (and for Sasol's original company town, Sasolburg) coalesced around the idea of 'respectable' and industrialised citizens (Sparks, 2012). Such 'model citizens' were also to subscribe to the ideals of modernity and, during Secunda's early years, to the ideologies of apartheid as based on outward appearance and certain behaviours.

As a result of the encouragement of 'respectable' citizens, the Sasol gym (which is in close proximity to the former 'Checkers Mall') added a sense of Western modernisation and 'respectability' to the general area, garnering comfort for some white people. Suna indicates a sense of not belonging in the 'Pick n Pay Mall', which translates into topophobia. It is also noteworthy that the 'area near Mr Price' has a more open design, which allows for greater visibility/'eyes on the street' which may lessen instances of crime (Jacobs, 1961).

This scenario highlights a distinct nuance separating topophobia from symbolic violence. It shows that, as a result of their topophobia, some people may resort to symbolic violence to maintain various status quos that uphold their privilege. For example, a topophobia of 'the racial other' may lead to a retreat from, or abandonment of, a place altogether.

In the case where Secunda's CBD is no longer solely a 'white' space, topophobia has been accompanied by symbolic violence in an attempt at regaining dominance. However, here, the use of symbolic violence has proven ineffective: despite attempts by some who long for a 'white people only' Secunda, the Secunda CBD has diversified and is no longer a 'white' space. In this sense, even though topophobia may have been heightened when dominance over space is threatened, attempts at symbolic violence are muted because they are no longer being supported by instruments of the State (for example the former Group Areas Act). However, in spaces where topophobia is supported by State instruments, such as many of Secunda's green belts that were fenced off during the private-public-partnership, Ikusasa, in an attempt to negate the effects of crime and decay (journal notes), symbolic violence can be used to bolster fears of 'the other'. Examples of this include the 'othering' of older children (as a whole) who are labelled as either peddling or using drugs when 'hanging out' in the green strips, or by linking a perceived increase in crime to an increase in black pedestrians who have been diverted through certain green strips through the limiting of access to others (journal notes).

This interview took place a few months before the opening of (what participants refer to as) 'the new mall' (opened in November 2013). Theerafter, all participants expressed a preference for the 'new mall':

> *The mall is one of the places in town I feel most safe – I don't mind coming here alone. I don't go to the post office alone anymore. We go on a Saturday when my husband can come along because I refuse to go there alone… That area just feels…although they say that they have changed the old Pick n Pay and that it's safe and pretty, but I still won't shop there.*
> *(Loraine, translated 26/03/2015)*

> *I don't really go to the town centre now that we have the mall.*
> *(Marlie, translated 30/04/2015)*

> *I have to go and gym [spaza caravans are located opposite, inter alia, here] and have to stop at that Spar then I don't feel safe… laughs. It feels a little bit run down and it feels isolated and thus I would much rather go to the new mall than stop at any of those shops… I think it's because there aren't so many [white] people that go to the shops anymore… Since the new mall has opened, I don't really feel safe in the old mall.*
> *(Corinne, translated 25/08/2015)*

These participants reveal that a dichotomous thinking, based on whiteness, between the 'old' and 'new' mall has manifested in the meaning of Secunda's space. The history of this dichotomy can be traced to original identities that the early residents of Secunda, who were mostly Sasol employees, were expected to aspire to, namely model, modern industrial citizens: An identity that was also infused with apartheid ideologies of whiteness (Sparks, 2012). In addition to this, the equation of 'modernisation' with 'Western' and 'development', says Sihlongonyane (2015: 67), is a product of Eurocentric thinking entrenched through powerful marketing strategies promoting the "Western city as an object of desire in South Africa". As such, instead of basing their identities on whiteness, says Ballard (2004: 55) many white people seek to assert a 'superiority' based on Eurocentric catch phrases such as "'developed', 'modern', 'Western', 'First World'".

References made by many participants to the old town or 'old mall' (*ou dorp*) indicate that some individuals no longer perceive the town centre to be a central cornerstone of the town (journal notes). For these participants, it has become 'old', 'not-modern', and 'undesirable'. A dichotomy between the old and new has been aggravated by a planning decision to locate the 'new mall' adjacent to the town centre. These dichotomies include 'working-class/middle-class', 'modern/old', and 'formalised-Western/informal- 'the other'' and are evidence of a symbolic barrier that has been developed to distance white residents from 'the racial other'. Arguably, these dichotomies may prove to be a direct hurdle to urban transformation in South Africa, particularly as "informality has become a dominant mode of urbanisation in much of the global South" (Sihlongonyane, 2015: 70).

Topophobias that were once ascribed to decay by some participants, remain – even though Resilient Properties, which also developed the 'new mall', has recently refurbished much of the town centre (journal notes). This reveals deep-seated topophobias of 'the racial other' which have come to the fore through a decrease in whiteness and loss of 'white' dominance over the town centre. Ballard (2004: 58) indicates that the influx of black people to former 'white' spaces has, in the views of some of his participants, altered it from a (European) space that "generated a sense of 'white' achievement" to one of few rules, or disorder: "Quite simply, home no longer feels very homely".

Further, argues DiAngelo (2011: 61), white people "often confuse comfort with safety". Participant topophobias of the 'loss' of dominance in former comfort zones may lead to attempts at the reassertion of dominance elsewhere (for example, in newly developed, Westernised spaces), or the avoidance of 'other' spaces. The effects of this on desegregation and social integration can be extremely damaging as "fear has an unholy alliance with prejudice, hate and anger" (Koskela, 2010: 390).

Notably, Sasol spearheaded the development of the 'new mall' as part of a bid to retain a skilled workforce (Bert, 14/05/2015; 15/05/2015). With the demise of apartheid and the adoption of a stronger capitalist focus, Sasol's paternalism is no longer as pronounced as it was previously (Sparks, 2012). However, Secunda remains a pseudo-company town, which Porteous (1970) says, experiences the indirect, unofficial control (as a dominant source of employment) by a corporation even though it is municipally administered. As such, Sasol is often portrayed as a 'hero' who 'saves' residents from what many participants perceive as an 'incompetent' municipality:

> *Sasol has the manpower [sic] to come and rescue.*
> *(Suna, translated 11/02/2013)*

> *I'm very disappointed in the way our town is backsliding. For example, one morning we were left without water because the municipality failed to pay the water council. Sasol had to pay that account. I must be honest with you, I think that if it wasn't for Sasol then our*

town would be much worse off. Go and have a look over weekends: you'll see Sasol employees cleaning the town. They wear the blue Sasol overalls but with yellow stickers on them.
(Estie, translated 31/03/2015)

Sasol has workers who maintain the sidewalks and pick up the rubbish. Sasol has even given the municipality money to fix the streetlights because the municipality doesn't have enough money to maintain everything… Sasol does the municipality's job for them.
(Melissa, translated 27/02/2015; 17/06/2015)

Within a new political environment, Afrikaners have increasingly become a 'racial other' and have experienced "[d]eep-seated anxieties about identity and loss of self" (Steyn, 2004: 153). Historically many Afrikaners have perceived themselves as being oppressed: First by a powerful English regime, and then in a post-apartheid era, by "African majority rulers" (Steyn, 2004: 159). As an ANC-run entity, Govan Mbeki Local Municipality (GMM), under whose jurisdiction Secunda now falls, represents the conquering 'racial other'. However, the municipality, like many others, has arguably entered into power carrying the heavy burden of what Sihlongonyane (2015: 69) calls an "inferiority complex" perpetuated by Western binaries, and colonial-apartheid dogma that attached identities such as 'unfaithful', 'disorderly' and 'incompetent' to 'the racial other' (Vestergaard, 2001).

For Estie, a sense of up-rootedness is aggravated by physical boundaries:

People used to be so proud of Secunda: I'm not sure what has happened. The palisade fencing that encircles the municipality – that was never there! There were no palisades and things around the library or post office.
(Estie, translated 31/03/2015)

The palisades, and by extension the current administration, represent danger for Estie: And might be perceived as a barricade by some. As symbols of municipal prowess, the aesthetic value of civic centres has a great impact on the topophobias of residents. Burns (2000) demonstrates that power forms are embodied by the spatial – such as prominent buildings (for example, the skyscraper) and centrally located facilities. The Secunda civic centre was designed to radiate power through its central location, a prominent group of buildings, and aesthetically pleasing street furniture. However, a civic centre that has been allowed to age goes against Western values of modernisation (see above), and can thus instil anxieties over the local administration's interest in its constituents: Decreasing trust and heightening the 'us and them' dichotomy between local officials and some white residents:

When we went to complain about it [an electrical fault], the officials were very difficult and very rude. If you look at the service that they deliver, Sasol has to keep the town tidy the municipality doesn't do it… So, if I had to change something, I would definitely change our municipality: Their attitude. If you have to go and get your licence… It's terrible, filthy. The municipality is the one thing that most residents complain about in the town.
(Sandra, translated 25/03/2015)

The challenges and competencies of local government, although an important topic, is beyond the scope of this chapter. What is important to this study is that a municipal-resident relationship can be weakened (inter alia), not only by stigmatisms inherited from colonial, apartheid, and modernist planning, but also by the topophobias resulting from these. In the case of Secunda, these topophobias have led some residents to cling to Sasol as a 'known' in what might have become an 'unknown' urban environment. Anecdotal research indicates that this diminished relationship could also be linked to the difficulty currently experienced by the municipality in its tax collection (journal notes). The next section will consider the use of symbolic boundaries, as a form of symbolic violence and a reaction to topophobia, to maintain Secunda as a 'white' comfort space.

2.5.3 The Symbolic Boundary of Market Filters

> *[I]f you want to stay in Secunda you pay a premium... Secunda residents pay the exact same percentage [tax] as the residents in eMbalenhle [Secunda's former 'black township'] but the value of property is much higher.*
> *(Greg, 07/05/2015)*

Greg, a former municipal official of Secunda, sets out a deep-seated disparity in the property values of the former black and white apartheid townships. The presence of a strong market filter in Secunda arguably acts as a symbolic barrier to desegregation in the town. As such, this assists in maintaining the town's former 'white' middle-class neighbourhoods as 'comfort zones', or enclaves of whiteness. Large disparities in property values can act as a symbolic barrier, limit mobility between former 'black' and 'white' spaces (and racial maintain segregation) and have assisted a shift from racial to class-based segregation in South Africa (Schensul & Heller, 2011). As such, market filters can be seen as a form of symbolic boundaries. Ballard (2004: 57) explains this in more detail saying that market filters are affected in many middle-class urban areas:

> *[W]hereby only 'desirable' people are able to move near 'whites' or take up occupation in cities because of the high cost of renting or buying property in these areas.*

Greg's (30/06/2015) perception highlights the presence of such a market filter (or symbolic boundary) in Secunda:

> *I've realised over the past years especially from 1995, you get an economically active black person who is more traditional who does not want to come and stay in Secunda: That person would prefer to stay in eMbalenhle. Go and look at the place that he [sic] is staying in – it is one of the most beautiful places [houses] that you can think of. But then, you get the other ['other'?] kind that is more Westernised. That person wants to have a different type of environment and can afford to change [areas]... [T]here is a small amount of your higher end income group of R1 million plus that you will find has moved to Secunda.*

Greg's perception reveals that racial desegregation in Secunda is mainly based on the presence of a more 'Westernised' 'the racial other' who can afford houses of R1 million and above, while 'other' more 'traditional' black people are perceived to prefer the former 'black township' of eMbalenhle. In the same interview, he indicates that individuals who can afford houses with values less than R500 000 are classified as "lower income groups" and "have moved to Evander" (an adjacent town – see Fig. 1 above). Entering into the Secunda market can be difficult due to a scarcity of developable land, high salaries paid by Sasol, and tight control held by local estate agents who have driven up the cost of properties (Greg, 30/06/2015). A blog spot reveals a conversation held by former and current residents of the town[5]:

> *My neighbour wants to develop a security complex of more affordable housing. He says there are a lot of young people in the town that can't afford today's [housing] prices. He says one of the guys working for him is hiring a small flat in someone's backyard for something like R4000. It's just a garage with a door: you know, not even a nice apartment.*
> *(Suzette, translated 15/01/2015)*

> *Rent is pretty steep for the sticks.*
> (Chris 2.0, 2008)

> *[T]he living conditions here aren't so bad, prices are steep though…*
> (Lonewolf, 2008)

> *My friend that I met here in Secunda from Pietermaritzburg offered me a place to stay until I get a place of my own. It's a huge mission to get a place here that fits my budget.*
> (Giken, 2009)

The above conversation indicates that Secunda Proper is still classified by predominantly uniform, low-density neighbourhoods (one house, one stand) (journal notes). This is especially the case for the new generation middle-class and the working-class (journal notes). Further, individual anecdotes have been recorded where black people have been discouraged from buying in Secunda's neighbourhood by white individuals (journal notes). These include, (i) an unwillingness to sell to black people by some white residents ('we couldn't do that to our neighbours'), and (ii) a hesitation by local estate agents to take a black engineer to view listings in the Green Area, recommending homes in Evander instead ("When I was buying my house in Secunda, estate agents didn't want to take me to viewings in the Green Area etc. and recommended Evander") (journal notes). However, it is acknowledged that greater research evidence is required here.

In addition, a continuation in the largely uniform housing typology of "low-density, largely single-storey sprawl" typical to the 'white' spaces (and nuclear family model) of urban apartheid (Sihlongonyane, 2002) is one example of a lack of transformation in Secunda. Neighbourhoods are predominately low in density (although the municipal SDF refers to this as 'medium density'[6]) with stand sizes ranging between 700–2000 m^2 and stand-alone houses with a minimum of three bedrooms. Although medium-low density accommodation is available (for example, walk-up blocks of flats and a few complexes), this housing typology is in the vast minority (Fig. 2.8).

---

[5] The handles related to these conversations (Chris 2.0, Lonewolf and Giken) were not changed as they are pseudonyms chosen by the authors who essentially wrote anonymously on a public domain.

[6] Here, residential densities are classified as: low density: 10 units/ha, medium: 30 units/ha, medium-high: 70 units/ha, and high 100: units/ha (Department of Rural Development and Land Reform, 2014).

TYPE OF USE	ERVEN N°	AREA m²	TOTAL AREA%
DETACHED DWELLINGS COMBINED DWELLINGS	1460 23	141,14 53,50	40,25 15,25
RESIDENTIAL (c)	1483	194,64	55,50
SECONDARY SCHOOLS PRIMARY SCHOOLS NURSERY SCHOOLS	1 2 2	8,82 7,00 0,47	2,52 2,00 0,13
EDUCATIONAL	5	16,29	4,65
PARKS TREEBELTS	29 9	38,56 11,70	10,99 3,34
OPEN SPACES	38	50,26	14,33
SHOPS & GARAGES	4	2,24	0,64
CHURCHES	12	4,34	1,24
INSTITUTIONS (b)	1	1,17	0,33
MAIN ROADS LOCAL STREETS		25,18 56,58	7,18 16,13
TRAFFIC		81,76	23,31
TOTAL AREA (a)	1543	350,70	100

Fig. 2.8 The original summary of uses for Secunda's first residential area. (Source: Kirchhofer (1982: 24))

Figure 2.8 indicates that 1460 stands for 'detached dwellings' were set out for Secunda's first neighbourhood, while only twenty-three 'combined dwellings' were made provision for. The current situation remains largely the same where 5388 single unit stands form a majority over medium-high density erven (1962) (Department of Rural Development and Land Reform, 2014). A uniform housing typology based on the nuclear family typical to the modernist planning during Secunda's infant stage (see above), cannot meet the needs of a more diverse society in Secunda. For example, a new generation middle-class (which is not necessarily based on the nuclear family) may not be in favour of, willing or able to take on the maintenance attached to a 1000 m² erf.

Higher density developments have recently begun to address the demand for more affordable housing stock, but are mostly located in the nearby town of Trichardt or around the town centre (in the form of blocks of flats) – which, as shown above, is no longer a 'white' comfort zone. According to www.property24.com, (currently, South Africa's leading property sales website), the Trichardt developments involve a number of gated estates with smaller erven of between 166 – 515 m². The municipal Spatial Development Framework (SDF) also sets out a proposed development, which seeks to integrate the towns of eMbalenhle, Evander and Secunda:

> *It is foreseen that future development in this area will consist mainly of mass subsidy-linked housing schemes and other types of residential development.*
> (Department of Rural Development and Land Reform, 2014: 330)

Further provision is made for infill development ("infill of open spaces including green strips") and a certain extent of densification ("the approval of more than one unit per property limited to densities of units per hectare") in Secunda's neighbourhoods (NM, email correspondence 11/01/2017). These proposed initiatives may one day contribute to a more integrated region. However, these proposals seem to largely encourage desegregation in zones (reminiscent of the ordering-through-zoning approach of modernist planning). For example, with higher density housing mainly in the town centre and in Trichardt; subsidised housing in corridors outside Secunda; and densification in Secunda's neighbourhoods that has been left mainly to the (currently white-dominated) private sector. Seemingly, the transformation initiatives for Secunda may largely fail to penetrate existing symbolic boundaries or alter the current *status quo* of dominance in its neighbourhoods.

Lemanski (2006) echoes a reluctance for the involvement of South Africa urban planning policies in social integration. She also argues that social mixing should be encouraged in public spaces only, "moving the focus of integration away from the neighbourhood" as doing so may infringe on the rights of some (Lemanski, 2006: 584). While it is acknowledged that the protection of human rights is fundamental, the case of Secunda serves to caution planners against generalised recommendations (especially when related to aspects such as a topophobias of 'the other'). Another form of symbolic boundary, namely attitudes of distancing, prevalent in Secunda's situated context will be considered next.

2.6 Ignorance and Symbolic Boundaries of Distancing: Perceptions of It's Not a Problem of 'Respectable' White People

As an inherited construction existing in the subconscious of beneficiaries, white privilege is not easy to grasp or acknowledge (Leonardo, 2004). The latency of white privilege is perhaps its most disturbing aspect when considering its potential for inflicting symbolic violence and contributing to latent topophobia. Because white privilege is often deeply ingrained in the subconscious of some white people, it may lead to an ignorance of the affliction (Sullivan, 2014). Sullivan (2014: 595) refers to the term white ignorance (a potential condition of white privilege) as:

> *[A] product of white people's unconscious racial habits, which have deep roots and are strongly invested—albeit not consciously—in maintaining the economic, psychological, and global domination of people of color.*

Through white ignorance, argues Hubbard (2005: 54), whiteness itself can become an "unnamed norm against which Otherness is gauged." Sullivan (2014) argues that white ignorance affects both educated and uneducated people. This often manifests itself as knowledge that reinforces a world in which white people feel comfortable socially, economically and psychologically (sometimes at the expense of other race groups), but can simultaneously render us blind to certain privileges and may also cripple our moral capacity to identify unethical behaviour (Sullivan, 2014).

Another functioning of white ignorance is that white people may 'ignore', or be unaware of, various racially-based privileges (Leonardo, 2004; DiAngelo, 2011). In this manner, certain white identities can distance themselves from racial inequality, which becomes a "problem of the other " (Leonardo, 2004: 143). This was highlighted one evening during a conversation on the high crime rate in South Africa (journal notes). One participant expressed an anxiety based on information that women in South Africa have a one in three chance of being raped. In an attempt to alleviate her concerns, another group member replied, "yes, but that happens mainly in the townships" (journal notes). For this participant, the concern related to such a high rape rate was reduced as it was less likely to apply to white, middle-class spaces. In other words, such a high rape rate was perceived as 'not a white problem'. This disassociation says DiAngelo (2011: 55), is not uncommon as problems of racial inequality are often ascribed to "what 'they' have, not us". The urban planning realm has also been criticised for similar disassociations evident in the use of phrases like 'urban', 'inner city', and 'disadvantaged' rather than 'white', 'over-advantaged' or 'privileged' (DiAngelo, 2011).

Another form of racially-based dissociation can be found in Rosie's account:

> *I sat on those benches [at a local pharmacy] they have for people waiting in the queue to collect their prescription tablets. An older white man, sitting on my one side, smiled at me which made me feel more positive about having to wait in a queue, which I hate doing. Two white women, with short hairstyles and tattoos, which looked strange when considering their age, approached us. When they saw me, they stood and openly talked about whether or not to take a place on the bench next to me. They eventually decided not to, as I was not white, and to return another time. It was embarrassing. How can this still be happening, how can individuals still be this way? When are they going to realise that times have changed?*[7]
>
> *(Rosie, Facebook update: September 2015)*

Rosie's post received over twenty-four comments, all of which were supportive. All except three were from both white and black women. Three comments, two in agreement with the first, advised Rosie to ignore them as *'it's only low-class whites that act that way'*. Some commentators indicated their shame at being white as it linked them to *'stupid and common (or 'commin') people like that'*, others blamed the two women as having no manners and another indicated that having such hairstyles and tattoos was a sign that they were 'not-respectable'.

[7] Note that Rosie's account was in English but her name has been changed and her comment rewritten to ensure confidentiality.

These responses indicate a distancing from 'bad' identities of whiteness, which are seen to contravene certain middle-class codes of 'respectability'. This stance seeks to protect the moral standing of certain whiteness identities by transferring culpability to localised individuals, thereby allowing certain white people to 'free' themselves from conviction (Leonardo, 2004; DiAngelo, 2011):

> *It must be the position of a good white person to declare that racism is always about 'other whites,' perhaps 'those working-class whites.' This is a general alibi to create the 'racist' as always other, the self being an exception.* Leonardo (2004: 144)

As a result of this distancing, few white people are able to identify certain actions as racially-fuelled (Leonardo, 2004). Distancing can have a harmful outcome in that it may recreate white privilege "despite good intentions" (Leonardo, 2004: 144), as it affords those white people who distance themselves the choice of "when, how, and how much to address or challenge racism" (DiAngelo, 2011: 64).

Further, Baum (2015) argues that rational planning largely views emotions as external to the rational 'self' and depends on an ignorance of emotions. Thus, viewing individuals (and planners), not as actors, but as passive objects to emotions, can lead to a release from the responsibility of 'emotionally-driven' actions as well as a lack of introspection of 'the self'. This could produce attitudes of 'it's not a problem of *respectable* white people' who are, by virtue of certain Western values, often expected to be rational. A lack of introspection, or ignorance of emotions of 'the self', could also heighten topophobias produced by the condition of instant blackness (where desires to be 'pro-black' or 'non-racist' may go hand-in-hand with the reinforcing of the *status quo* of white privilege and dominance). It could also strike cords of white fragility (DiAngelo, 2011), missing opportunities for constructive interaction about issues of race, and possibly lead to the (symbolically) violent responses typical to white fragility, for example, argumentation, silence, or a distancing from stressful (engaging) environments.

A second outcome of viewing emotions as external is that this view may overlook an individual's responsibility for their emotional reactions (Baum, 2015). Through a lack of introspection, actions can become defensive and/or controlling – or, as shown above, take on a form of distancing where racially-based urban issues are 'not a problem of white people', even though they may be caused by systems of symbolic violence such as white privilege.

2.7 Conclusion

This chapter considered the presence of symbolic violence as a reaction to topophobias related to maintaining whiteness within Secunda. Specifically, it focussed on the symbolic boundaries of white privilege and how transformation can be compromised – both in perceived and lived space – to maintain enclaves of whiteness. The increasing presence of 'the racial other' has led to various anxieties for certain whiteness identities. This has amplified the meting out of symbolic violence by some white people in an attempt to maintain dominance. Although this dominance has been lessened in more desegregation spaces, such as the town centre, Secunda remains largely an enclave for white privilege.

The weakening of white dominance has resulted in dichotomous thinking by some participants, who saw the town centre as 'old', 'not-modern', and 'disorderly'. Dichotomies serve to segregate the town centre from the nearby 'new mall', the middle-class from the working, formalised from informal, and ultimately, 'the same' from 'the other'. These dichotomies can be a direct hurdle to transformation in South Africa, especially as informality is a dominant (and little-understood) form of urbanisation in many African spaces.

Two main comfort enclaves were identified and explored in Secunda, namely the Duck Pond and the CBD. Initially, the Duck Pond was the apex of Secunda's green strips (which symbolised white dominance and privilege – above). With the nullification of the Group Areas Act, the Duck Pond has benefited from a diversity of users – both resident and out-of-town visitors. This disruption of white dominance resulted in a perceived undermining of identity for some of my participants leading them to a struggle for space – or, in some cases – the abandonment of these former comfort enclaves. Similarly, a consideration of the topophobias that have arisen from the introduction of informal businesses or 'spaza' shops into the CBD shows how symbolic violence is meted out in an attempt to reinforce symbolic barriers between the racial other and 'the white/Western same'.

Another consequence of a loss of dominance in certain former 'white'-dominated spaces in Secunda is a turning to Sasol as a 'hero' in a new, or 'unknown', environment. Upgrading initiatives and various maintenance activities run by the company (in continued interests of retaining a skilled workforce) retain Sasol's 'hero' status, even though the company's focus has taken a more capitalistic stance, making its paternalism no longer as pronounced.

Two other forms of symbolic boundaries were considered, namely market filters and attitudes of distancing, which hinder transformation in Secunda. Market filters act as invisible barriers to certain individuals who cannot afford the property prices of an area. This has become prevalent in a South African society, which is becoming increasingly class-based. Market filters in Secunda are mostly upheld by a scarcity of developable land, predominantly low density, uniform neighbourhoods, and high property values that are inflated by comparatively large salaries paid by Sasol and a tight market control held by local estate agents.

The symbolic violence and topophobias of a place can be so deeply entrenched that they become part of the everyday. When this happens, they are embedded in the subconscious and may or may not have reactions that are intentionally violent. Arguably, in the absence of state-condoned boundaries and considering the nature of symbolic boundaries, which is usually to intimidate and humiliate, symbolic boundaries are most used in the meting out of symbolic violence today.

There is no neatly packaged answer in the journey of overcoming white privilege. In order to fully comprehend the complexities of South African urban space, planners need to recognise and understand the spatial consequences of white privilege. Topophobia, with its accompanying pervasive anxieties and symbolic violence, is only one of these consequences. In making decisions on behalf of greater society, great value is to be found in confronting personal emotions and subjectivities – a resistance to do so has had, and will continue, to have debilitating consequences for real social and urban transformation.

References

Baeten, G. (2002). Hypochondriac geographies of the city and the new urban dystopia. *City, 6*(1), 103–115.

Ballard, R. (2004). Assimilation, emigration, semigration, and integration: 'white' peoples' strategies for finding a comfort zone in post-apartheid South Africa. In N. Distiller & M. Steyn (Eds.), *Under construction: 'Race' and identity in South Africa today* (pp. 51–66). Heinemann.

Baum, H. (2015). Planning with half a mind: Why planners resist emotion. *Planning Theory & Practice, 16*(4), 498–516.

Bauman, Z. (1992). Introduction: The re-enchantment of the world, or, how can one narrate post-modernity? In *Intimations of postmodernity*. Routledge.

Baxter, J. E. (2012). The paradox of a Capitalist Utopia: Visionary ideals and lived experience in the Pullman community 1880–1900. *International Journal of Historical Archaeology, 16*, 651–665.

Bremner, J. D. (2004). Bounded spaces: Demographic anxieties in post-apartheid Johannesburg. *Social Identities, 10*(4), 455–468.

Burns, A. (2000). Emotion and urban experience: Implications for design. *Design Issues, 16*(3), 67–79.

Chris 2.0. (2008, August 24). A question for those lives in Secunda. *Broadband.*

Christopher, A. (1983). From Flint to Soweto: reflections on the colonial origins of the apartheid city. *Area, 15*(2), 145–149.

Davies, R. J. (1981). The spatial formation of the South African city. *GeoJournal, 2*(2), 59–72.

Department of Rural Development and Land Reform. (2014). *Govan Mbeki spatial development framework 2014–2034*.

DiAngelo, R. (2011). White fragility. *The International Journal of Critical Pedagogy, 3*(3), 54–70.

Durrheim, K., & Dixon, J. (2010). Racial contact and change in South Africa. *Journal of Social Issues, 66*(2), 273–288.

Flyvbjerg, B. (2001). The Power of Example. In *Making social science matter* (pp. 66–87). Cambridge University Press.

Freund, B. (2013). *Social order and organisation in Vanderbijlpark and Sasolburg: The company towns of the Vaal Triangle 1940-901*. https://aehworkshop.files.wordpress.com

Friedmann, J. (1987). *Planning in the public domain*. Princeton University Press.

Friedmann, J. (2010). Place and place-making in cities: A global perspective. *Planning Theory & Practice, 11*(2), 149–165.

Giken, O. (2009, February 6). *My housemate doesn't want me there!*

Greed, C. (1994). The place of ethnography in planning: Or is it "real research"? *Planning Practice & Research, 9*(2), 119–128.

Gustin, G. (2010, September 26). *Food trucks bring haute cuisine and problems*. St. Louis Post-Dispatch.

Harrison, P., Todes, A., & Watson, V. (2008). *Planning and transformation: learning from the post-apartheid experience*. Routledge.

Hubbard, P. (2005). Accommodating otherness: Anti-asylum centre protest and the maintenance of white privilege. *Transactions of the Institute of British Geographers, 30*(1), 52–65.

Jacobs, J. (1961). *The death and life of great American Cities: The failure of town planning*. Random House.

Janz, B. (2008). The terror of the place: Anxieties of place and the cultural narrative of terrorism. *Ethics, Place & Environment, 11*(2), 191–203.

John, L. (2012). Secondary Cities in South Africa: The start of a conversation. South African Cities Network. Available: http://led.co.za/documents/secondary-cities-in-south-africa-the-start-of-a-conversation [2014, February 25].

Kirchhofer, M. (1982). The planning of Sasolburg and Secunda – Achievements and prospects. *Journal for Town and Regional Planning, 14*, 1–28.

Koskela, H. (2010). Fear and its others. In S. Smith, R. Pain, S. Marston, & J. Jones (Eds.), *The SAGE handbook of social geographies* (pp. 389–407). SAGE Publications.

"Latent, adj". (2014, February 25). *OED online*. Oxford University Press. www.oxforddictionaries.com

Lefebvre, H. (1971). *Everyday life in the modern world*. (Sacha Rabinovitch, Trans.). Penguin Press.

Lefebvre, H. (1991). *The production of space*. Blackwell Publishing.

Lemanski, C. L. (2004). A new apartheid? The spatial implications of fear of crime in Cape Town, South Africa. *Environment and Urbanization., 16*(2), 101–112.

Lemanski, C. L. (2006). Desegregation and integration as linked or distinct? Evidence from a previously 'white' suburb in post-apartheid Cape Town. *International Journal of Urban and Regional Research, 30*(3), 564–586.

Leonardo, Z. (2004). The color of supremacy: Beyond the discourse of "white privilege". *Educational Philosophy and Theory, 36*(2), 137–152.

Lonewolf. (2008, August 24). *A question for those lives in Secunda*. MyBroadband.

Mabin, A., & Smit, D. (1997). Reconstructing South Africa's cities? The making of urban planning 1900–2000. *Planning Perspectives, 12*, 193–223.

Meintjies, J. (1975). *Sasol 1950–1975*. Tafelberg.

Meth, P. (2016). Informal housing, gender, crime and violence: the role of design in urban South Africa. *British Journal of Criminology, 57*, 402–421.

Newton, C., & Schuermans, N. (2013). More than twenty years after the repeal of the Group Areas Act: Housing, spatial planning and urban development in post-apartheid South Africa. *Journal of Housing and the Built Environment, 28*(4), 579–587.

Oranje, M. (1998). *The language game of South African urban and regional planning: A cognitive mapping from the past into the future*. [Doctoral dissertation]. University of Pretoria.

Porteous, J. (1970). The nature of the company town. *Transactions of the Institute of British Geographers, 51*, 127–142.

Rees, A. (2012). Nineteenth-century planned industrial communities and the role of aesthetics in spatial practices: The visual ideologies of Pullman and Portsunlight. *Journal of Cultural Geography, 29*(2), 185–214.

Sandercock, L. (2005). Difference, fear and habitus: A political economy of urban fears. In *Habitus: A sense of place* (pp. 219–234). Ashgate.

Schensul, D., & Heller, P. (2011). Legacies, change and transformation in the post-apartheid City: Towards an urban sociological cartography. *International Journal of Urban and Regional Research, 35*(1), 78–109.

Sihlongonyane, M. F. (2002). Urban planning in the post-apartheid South Africa: Reconstruction and transformation. In A. Osmanovic (Ed.), *Transforming South Africa*. Institute of African Affairs.

Sihlongonyane, M. F. (2015). The challenges of theorising about the Global South a view from an African perspective. *Africa Insight, 45*(2), 59–74.

Sparks, S. J. (2012). *Apartheid modern: South Africa's Oil from coal project and the history of a South African Company Town*. [Doctoral dissertation]. The University of Michigan.

Spinks, C. (2001). *A new apartheid? Urban spatiality, fear of crime, and segregation in Cape Town, South Africa*. Working Document. London School of Economics and Political Science, Development Studies Institute. Available: http://www.lse.ac.uk/internationaldevelopment/pdf/wp/wp20

Stacey, J. (1991). Can there be a feminist ethnography? In D. Gluck & S. B. Patai (Eds.), *Women's words: The feminist practice of oral history* (pp. 111–119). Routledge.
StatsSA. (2001, May 2015). *Population census 2001*. www.statssa.gov.za
StatsSA. (2011, May 25). *Population census 2011*. www.statssa.gov.za
Steyn, M. (2004). Rehabilitating a whiteness disgraced: Afrikaner white talk in post-apartheid South Africa. *Communication Quarterly, 52*(2), 143–169.
Steyn, M., & Foster, D. (2008). Repertoires for talking white: Resistant whiteness in post-apartheid South Africa. *Ethnic and Racial Studies, 31*(1), 25–51.
Sullivan, S. (2014). The hearts and guts of white people. *Journal of Religious Ethics, 42*(4), 591–611.
Thomas, R. (2003). *Blending qualitative and quantitative research methods in theses and dissertations*. Corwin Press.
Tuan, Y. (1974). *Topophilia: A study of environmental perception, attitudes, and values*. Prentice-Hall.
Tuan, Y. (1979). *Landscapes of fear*. Basil Blackwell Publisher.
Vestergaard, M. (2001). Why South Africa matters. *Daedalus, 130*(1), 19–44.
Visser, G. (2001). Social justice, integrated development planning and post-apartheid urban reconstruction. *Urban Studies, 38*(10), 1673–1699.
Watson, V. (2013). Planning and the 'stubborn realities' of global south-east cities: Some emerging ideas. *Planning Theory, 12*(1), 81–100.
Wessels, P. (1990). *Crescendo to Success: Sasol 1975–1987*. Human & Rousseau.
Whitzman, C. (2007). Stuck at the front door: Gender, fear of crime and the challenge of creating safer space. *Environment and Planning A, 39*(11), 2715–2732.
Yiftachel, O., & Yacobi, H. (2003). Urban ethnocracy: Ethnicization and the production of space in an Israeli "mixed city". *Environment and Planning D: Society and Space, 21*(6), 673–693.

Sketch Artists

Al-Najar, Z. (2023). *Making amagwinya* [Mixed Medium]. Architectural Student, Qatar University.
Elbadri, N. (2023). *A sketch of the Spaza Caravans Llocated in the Secunda CBD* [Mixed Medium]. Architectural Student, Qatar University.

Chapter 3
Neglect of Municipal Infrastructure in Phalaborwa

Manfred Spocter

3.1 Introduction

The construction and maintenance of infrastructure in the built environment is essential for promoting economic growth and ensuring that people have access to basic services. Conversely, the neglect of infrastructure negatively impacts opportunities for economic growth and the provision of basic services. Local government plays a key role in infrastructure and basic service delivery, a daunting task for many local municipalities in South Africa. This chapter investigates the state of the municipal infrastructure in the town of Phalaborwa, Limpopo province.

The evidence presented in this chapter shows that the Ba-Phalaborwa Local Municipality (LM) has not been honouring its mandate of sustainable and affordable service delivery because of years of neglect of and underinvestment in infrastructure. This state of affairs adversely affects its citizens, who have for years tried to engage the municipality to fulfil its mandate, to no avail. All the interviewees in the study were of the opinion that the infrastructure related to potable water, electricity, wastewater, solid waste, roads, parks and open spaces in Phalaborwa are in dire need of cleaning, refurbishment, upgrading, renewal, expansion and replacement. Infrastructural failures are characterised by disruptions of water and electricity supply without prior warning, toppling streetlights, leaking wastewater pipes and the nauseating odour of raw sewerage in watercourses. Hence, in addition to business initiatives, this chapter spotlights the efforts of various past and present residents' associations to turn the situation around.

Data for the study was collected from a plethora of municipal documents, government, consultant and non-profit organisation reports, academic literature and

M. Spocter (✉)
Department of Geography and Environmental Studies, Stellenbosch University, Stellenbosch, South Africa
e-mail: mspocter@sun.ac.za

© The Author(s), under exclusive license to Springer Nature Switzerland AG 2023

R. Donaldson (ed.), *Socio-Spatial Small Town Dynamics in South Africa*, GeoJournal Library, https://doi.org/10.1007/978-3-031-37142-4_3

online video clips. Personal semi-structured interviews were conducted with current and retired senior municipal officials, the corporate affairs and stakeholder engagement managers of Foskor and the Palabora Mining Company (PMC), a representative of the Ba-Phalaborwa Tourism Association, the steering committee of the Let's Change Ba-Phalaborwa (LCBP) forum and the then chairperson of the defunct ratepayers' association. Views on particular aspects of the study were also sought from a local economic development (LED) practitioner and a qualified electrical engineer.

After this introductory section, the involvement of civil society in attempts to improve the state of municipal infrastructure in South Africa is presented followed by an overview of the historic development of Phalaborwa. Thereafter, municipal infrastructure in Phalaborwa is covered, focusing on potable water, wastewater, electricity, solid waste, and roads, parks and open spaces. Financial challenges faced by the municipality are also presented. The support of the local mining companies in addressing municipal service delivery issues is spotlighted as well as the involvement of civil society. The chapter concludes with some parting thoughts.

3.2 Civil Society and Municipal Service Delivery and Infrastructure in South Africa

The parlous state of municipal service delivery, which includes the upkeep of municipal infrastructure, throughout the country is briefly reviewed in this section. The infrastructure problems experienced in the Ba-Phalaborwa LM is not an uncommon state of affairs and mirrors the challenges faced by many local municipalities in South Africa. The differences lie in how the infrastructure challenges are addressed, the action and non-action of civil society and businesses, and the willingness, or lack thereof, of municipalities to accept assistance and work together with citizens' and business groups.

Financial issues are at the heart of municipal infrastructure challenges. Reporting on the financial position of the 257 municipalities in South Africa,[1] the National Treasury (2022) indicated that 175 municipalities (68%) were in financial distress with 151 (59%) listed as bankrupt and insolvent and 43 (17%) quagmired in a financial crisis. A staggering 219 municipalities (85%) are triggering Section 138 (criteria for determining serious financial problems) and Section 140 (criteria for determining serious or persistent material breach of financial commitments) of the Municipal Finance Management Act (Act 56 of 2003) which suggests an imminent service delivery failure (Republic of South Africa, 2003; National Treasury, 2022; SANews, 2022). This is attributed largely to ineffective municipal revenue collection and the failure of municipalities to pay for services received. The grim reality

[1] Comprising eight metropolitan municipalities, 44 district municipalities and 205 local municipalities.

is that municipalities are not meeting their legislated mandate regarding service delivery and the maintenance of the current municipal infrastructure.

Civil society organisations in South Africa have a rich history of activist involvement in local-level issues (Weideman, 2015). While it is acknowledged that 'civil society' is a diverse concept, Beauclerk et al. (2011) explore the concept by defining five functions that it fulfils, namely, "… helping generate the social basis for democracy, promoting political accountability, producing social trust, reciprocity, and networks, supporting the rights of citizens and the concept of citizenship, and creating and promoting alternatives" (Taylor, 2013, p. 620). Furthermore, Boraine (2014, p. 126) writes that "… civil society has a very important role to play in emphasising the principles of democracy, transparency, integrity and justice …". Hence, there are numerous examples in towns in South Africa where citizens groups have been launched to tackle service delivery issues and failing infrastructure, through various means.

Civil society organisations are launching municipal-improvement actions separate from government (Venter, 2022). For example, in the town Secunda in Mpumalanga province, residents and businesses have taken it upon themselves to maintain road verges, paint road markings, clean parks and have launched an 'adopt-a-pothole' project (Bird, 2021). Likewise, the towns of Senekal and Modimolle, in the Free State and Limpopo provinces respectively, have witnessed similar citizen-driven initiatives (The Post, 2018; Lindeque, 2021). In Harrismith, also in the Free State, a citizen volunteer group, started in 2018, which funded and fixed water supply issues, electricity problems and repaired damaged infrastructure was stonewalled by the local government, resulting in the group contesting four wards as independent candidates in the 2021 local government elections – albeit without success (Karrim, 2021). In the face of countrywide service delivery failures and infrastructure challenges, civil society groups are mobilising in attempts to step in where municipalities have failed, as is happening in Phalaborwa. Before focusing on municipal service delivery in Phalaborwa an overview of the historic development of the town is presented.

3.3 Historic Development of Phalaborwa

The area in which the town of Phalaborwa is situated has been inextricably linked to mining activities for hundreds of years. Copper and iron resources were exploited in the Phalaborwa area in the recent and the deeper archaeological past and the oldest recorded copper mine in southern Africa is located there (Moffett, 2017; Moffett et al., 2020). Radiocarbon dating has determined the presence of mining activity at Selolwe hill, the site of an extinct volcano and the present-day PMC mine, that goes back to AD770 (More, 1974; Van der Merwe & Scully, 1971). Further archaeological evidence of pre-colonial extractive mining is present at three Early Iron Age sites in and around Phalaborwa, namely Kgopolwe, Nagome and Shankare. Kgopolwe Hill, a provincial heritage site associated with Kgopolwe ceramics of the

AD 900–1300 period, is located in Phalaborwa as is the Sealeng Hill on the outskirts of the town (Evers & Van der Merwe, 1987; Moffett, 2017; Republic of South Africa, 1970).

Over time, the area was settled by, inter alia, Venda, Tsonga, Sotho and Lobedu peoples. The oral traditions of the BaPhalaborwa[2]indicate that the settlement history of Phalaborwa is marked by the migration, interaction and assimilation of multiple communities (Moffett, 2017). Phalaborwa was a key point on the various trade routes (Delius et al., 2014), as recognised through historical footpaths and archaeological evidence, that connected to among others, the Mozambican coast. Iron and copper were important in interregional exchange along these trade routes and, as the area was not well-suited to pastoral and agricultural pursuits, the trade in metal goods was necessary to meet the subsistence requirements of the BaPhalaborwa (Delius, 1984). Specialised metal production in Phalaborwa continued until the nineteenth century; the "[m]etal resources sustained local communities and fed into local, regional and global exchange networks" (Moffett, 2017, p. 8). The mining of metal declined sharply as a number of factors, including increased European colonisation and increasing use of copper alloys from Europe, resulted in the cessation of indigenous copper mining and smelting and the collapse of the extraction-based economy by the 1880s (Miller, 1995; Plug, 1996; Killick et al., 2016). The indigenous people were dispossessed of their land as farms were surveyed and laid out, and they were forcibly relocated to small portions of land for residential purposes (Moroaswi, 2013). Until the 1950s the BaPhalaborwa were small-scale maize and millet farmers, with some cattle and small stock (Scully, 1978). The BaPhalaborwa were on the brink of relocation, as requested by the chief of the time because of an ongoing drought, when fortuitously the new commercial mining operations provided employment opportunities and the relocation plans were shelved (Van der Merwe & Scully, 1971).

The prospecting exploits during the first half of the twentieth century of Edward Mellor, Hans Merensky and others identified and confirmed the presence of phosphate and other minerals in the area. The importation of phosphate was halted by World War II and shortly after its cessation the then government recognised the need to reduce the country's dependency on imported phosphate fertiliser. The establishment of the town was predicated on the decision to commercially extract the rich mineral deposits. In 1951 the *Fosfaat-Ontginningskorporasie (Eiendoms) Beperk* [Phosphate Development Corporation (Proprietary) Limited] (Foskor for short) parastatal was registered with the mandate to mine the high-quality phosphate at Phalaborwa. Construction of the Foskor plant began in April 1952 and it was commissioned on 27 August 1955 (Cartwright, 1972, 1986; Roux et al., 1989).

[2]The Sotho prefix "Ba" (the) is a reference to the people of Phalaborwa. The name Phalaborwa means 'better than the south' because in the past the area was seen as a more desirable place to live than the lands to the south in present-day Bushbuckridge (Scully, 1978; Pottinger, 1990).

3.4 Phalaborwa: 1955 to Present

"If there was no Foskor, there would have been no Phalaborwa" (Pers. com., Foskor respondent, 2022); a statement that rings true as the town was established as a result of the commercial mining activities. Phalaborwa was founded to meet the need for housing for the white mine employees, but the exorbitant cost of building houses was beyond the means of the newly established Foskor. The National Housing Commission could not lend money to companies or individuals; thus, a local administration was established to access the finance. On 24 August 1955 the health committee of Phalaborwa was established, and this signalled the birth of the town. The health committee would preside over the affairs in the town for the next 14 years. The first 30 houses were built with a loan from the National Housing and Planning Commission and Foskor supplied the water, electricity, sanitation and sewerage services. PMC was registered in 1956, but mining and processing operations only commenced in 1965 (Pottinger, 1990). Phalaborwa was officially proclaimed a town in 1957 and a township for 10,000 Black employees was laid out beyond the boundaries of the town. A town council was proclaimed on 1 October 1969 (Grewar, 1980).

During apartheid, Phalaborwa was the white town and Black inhabitants were forced to reside in the townships of Namakgale and Lulekani, approximately 10 km from Phalaborwa, in the then Bantustans of Lebowa and Gazankulu, respectively. The dormitory townships were built in 1962 to house mine employees and their families (Ward, 2012). PMC financed the construction of 560 houses in Namakgale for their employees after 1972, the year in which Lebowa was declared a self-governing area. A further 80 houses, jointly financed by the Lebowa Development Trust and PMC, were built in 1978 and the two entities played a substantial role in the improvement and renovation of 1356 government-built houses. PMC also contributed to the construction of a 205-bed hospital and financed a sports stadium, a recreation club, a clinic and an administration centre, and they had the main roads tarred (Grewar, 1980; Pottinger, 1990). Foskor financed the construction of houses for its employees in Namakgale and Lulekani (Pers. com., Foskor respondent, 2022). Mine employees received title deeds to their homes from 1987 onwards as the government and the mining companies withdrew from schemes to provide housing. The mines launched social responsibility programmes in the late 1980s and established projects in partnership with the communities (Pottinger, 1990).

Transitional Local Councils (TLCs) were established countrywide in 1994 which constituted combined structures for, inter alia, former white towns, Black townships and traditional rural areas. The process culminated in the adoption of a wall-to-wall system of local municipalities throughout the country in 2000. The Ba-Phalaborwa LM is one of five local municipalities forming part of the Mopani District Municipality which has its council offices in Giyani 110 kilometres from Phalaborwa. There are four towns in the Ba-Phalaborwa LM, namely, Phalaborwa, Namakgale, Lulekani (see Fig. 3.1) and Gravelotte, together with 35 villages and 1029 farms. There are also five traditional authorities in the LM, namely the Mashishimale Traditional Authority (TA), the Maseke TA, the Majeje TA, the Ba-Phalaborwa TA and the Selwane TA. The population of the Ba-Phalaborwa LM is approximately

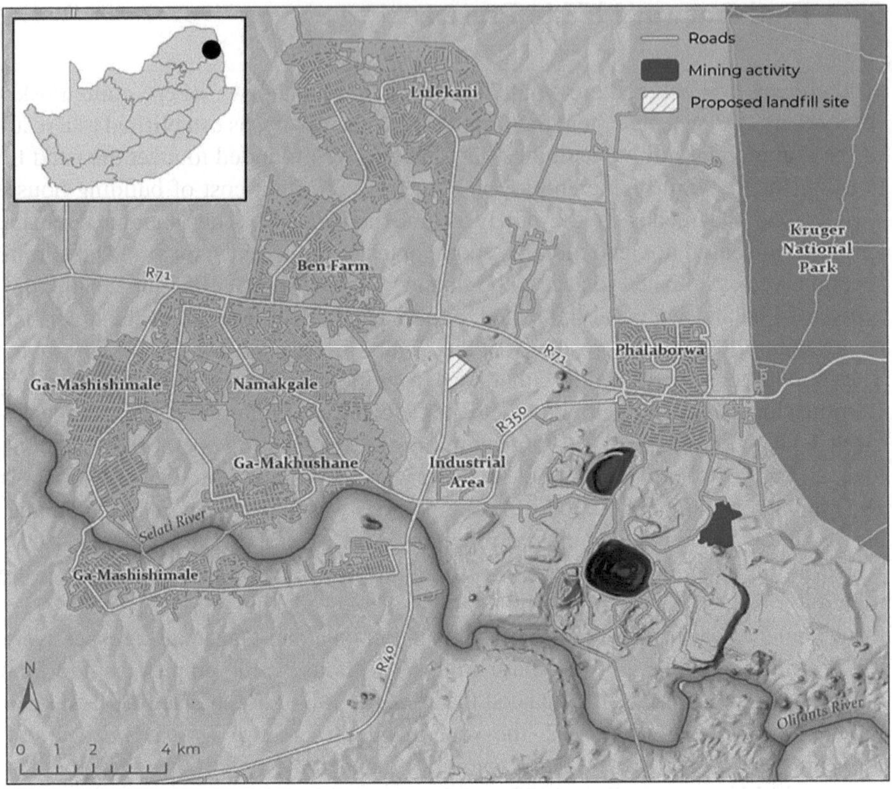

Fig. 3.1 Map of Phalaborwa and surrounding towns in 2022. (Source: Author (2022))

170,000 people with the town being home to around 32,000 residents. It is noteworthy that 94% of the municipality's population lives within a 15-kilometre radius of town (Ba-Phalaborwa Municipality, 2022a). Phalaborwa is inextricably linked to the nearby towns of Namakgale and Lulekani, the surrounding villages and the TA areas – spatially, economically, politically, socially and infrastructurally.

The economy of the Ba-Phalaborwa LM rests heavily on mining and tourism, with Phalaborwa being identified as a provincial growth point (Ba-Phalaborwa Municipality, 2013). There are two mines in close proximity to Phalaborwa. PMC extracts and beneficiates copper and vermiculite, and produces the by-products magnetite, sulphuric acid, anode slimes and nickel sulphate. The expected lifespan of the mine is 13 and 15 years for vermiculite and copper respectively, and 30 years for magnetite (Palabora Mining Company, n.d.). Phosphate is mined at the Foskor mine, with copper and magnetite as by-products (Pers. com., Foskor respondent, 2022). Rare earth elements are extracted from the gypsum stockpiles resulting from decades of phosphate mining (Bulbulia, 2020).

Tourism and mining seldom coexist in the same space, but Phalaborwa's location on the border of the world-renowned Kruger National Park (KNP) forces a coexistence of the two economic sectors. Phalaborwa is the only town situated directly adjacent to the KNP, with the Phalaborwa Gate one of the park's entry points and gateway to the northern parts of the park. The town has a host of tourism accommodation establishments used as a base for nature tourism (the KNP and surrounding game farms) and business tourism (related to the mines). Sixty-one percent of the land area in the Ba-Phalaborwa LM comprises game farms and 31% is devoted to nature- and conservation-related land use (Ba-Phalaborwa Municipality, 2013). The importance of the contribution of tourism to the local economy is recognised by the municipality in its vision that proclaims the "[p]rovision of quality services for community well-being and tourism development" (Ba-Phalaborwa Municipality, 2022a, p. 4) to be paramount, with one of the objectives being that Ba-Phalaborwa LM be positioned as the destination of choice (Ba-Phalaborwa Municipality, 2013). In fact, the 2009/2010 vision statement in the integrated development plan (IDP) read as "Best tourist destination in Limpopo by 2020" (Ba-Phalaborwa Municipality, 2011, p. 171). Tourism is envisioned as a major economic contributor in the post-mining local economy. However, for the tourism sector to flourish the municipality must provide the necessary infrastructure to support it. With this section as background, the scene is set to present current state of the municipal infrastructure in the Ba-Phalaborwa LM which is treated next.

3.5 The State of Municipal Infrastructure in Phalaborwa

This section overviews the state of infrastructure underlying certain municipal service delivery sectors in Phalaborwa. The intention is to provide the background and set the scene for the discussion of the involvement of businesses and civil society in inducing the local municipality to fulfil its service delivery mandate. Respondents opined that the infrastructure in town is in a bad condition due to a combination of factors: lack of human capacity, financial constraints, appointment of municipal staff unsuited to particular positions, administrative challenges, lack of accountability, failure of governance, cadre deployment and a lack of maintenance planning and foresight over the years. These causal factors are compounded when considering the population growth of Phalaborwa and the local municipality as a whole. The areal size of the town has not grown as it is hemmed in by the KNP in the east, the mining areas in the south, farms to the north and private property to the west. Instead, the town has densified in terms of buildings and population thereby increasing loads on the infrastructure which has not had any concomitant capacity increases (Pers. com., LCBP, 2022). Despite this, for one respondent the only expectation is that the local municipality should do what they are supposed to do; "deliver on their service delivery mandate, nothing else" (Pers. com., Tourism Association respondent, 2022).

The turnover of senior municipal staff, the changing priorities of politicians and the failure to delink politics from municipal administration have also been cited as factors contributing to shifting foci in service delivery by the local municipality. A tourism development strategy devised in 2006 prescribed that the roads, electricity, water supply and sanitation needed to be upgraded and that public amenities needed to be maintained, all in support of tourism (Ba-Phalaborwa Municipality, 2013). However, the most recent IDP identifies the upgrading of electricity infrastructure and sewer systems, the rehabilitation of streets and the low water pressure among the top five priorities in town (Ba-Phalaborwa Municipality, 2022a). The following subsections spotlight specific infrastructural challenges.

3.5.1 Potable Water Provision and Wastewater Management

The challenge that justifiably presents the most serious ongoing concern is the provision of potable water to Phalaborwa as it is indispensable and underpins all life; nothing survives without it. The Ba-Phalaborwa LM is the provider of retail water services and is ultimately responsible for water services to the town. The water services authority is the Mopani District Municipality (DM), while Lepelle Northern Water is the public water entity that provides bulk water services and reports to the national Department of Water and Sanitation (Lepelle Northern Water, 2021; Ndlozi, 2010). This situation of the local municipality not being a water services authority has been identified as a major obstacle to water delivery because of the fragmentation of administration and lack of coordination between the local municipality and the district municipality regarding this vital government service (Ba-Phalaborwa Municipality, 2013). The service level agreement (SLA) between the three different entities requires redrafting to ensure improved coordination, decision-making and service provision.

Although the water supply system to town is running at full capacity, the 2012 IDP divulged that the water distribution pipelines were in dire need of replacement as they were more than 41 years old, a decade ago. The ageing pipeline network is prone to breakages, leading to the loss of valuable water, disruptions of water supply and low water pressure. The local municipality is responsible for repairing and maintaining the potable water infrastructure, but it does not have up-to-date information on the amount of water loss. The unaccounted-for water adds to the tariffs the local municipality pays Lepelle Northern Water for bulk water (Ba-Phalaborwa Municipality, 2013, 2021a, 2022a, b). Water losses are high as a result of many leaks as well as an unknown number of illegal water connections. Many of these connections are attributed to the traditional authorities' allocation of erven[3] where

[3]The Spatial Planning and Land Use Management Act (Act 16 of 2013) requires the traditional authority to liaise with the municipality on any land use application involving land under its jurisdiction: "…the municipality must reach social compacts with the traditional authorities for the orderly development of their land…" (Ba-Phalaborwa Municipality, 2019a, p. 61). Contestations between local municipalities and traditional authorities about land use have been witnessed in the North-West, KwaZulu-Natal, Mpumalanga and Northern Cape provinces (Drummond & Nel, 2021). Simelane and Sihlongonyane (2021) propose that municipal and traditional authorities be "…an integrated and intertwined structure of urban governance" (p. 167).

bulk services infrastructure does not exist, so resulting in the residents connecting to the water lines illegally (Ba-Phalaborwa Municipality, 2019a). An advisable way forward would be to legalise the connections, install water meters and bill the residents for their water usage, following the granting of permission for such an action by the district municipality as per the SLA and by the traditional authorities (Pers. com., Senior municipal official, 2022). The current billing situation is patently unsustainable.

Despite the problems of cost recovery and unfinished investigations by the Special Investigating Unit[4] and the then Department of Human Settlements, Water and Sanitation, Lepelle Northern Water has a refurbishment scheme under way at Phalaborwa to increase the delivery of bulk water supply to the Phalaborwa industrial complex and to residential water users (Lepelle Northern Water, 2021). PMC's mining operations are not really impacted by inadequate water supply as the mines are the first to receive water from Lepelle Northern Water, even before residential consumers (Pers. com., PMC respondent, 2022). However, Foskor's greatest trouble concerns water supply, as the bulk of their water supply comes from the local municipality. In response, they have constructed a water recycling plant which reuses water for mining purposes (Pers. com., Foskor respondent, 2022). The upgrading of the bulk water supply infrastructure is being addressed by Lepelle Northern Water, but there is room to improve on other facets of the operation too.

The Blue Drop Certification programme administered by the national Department of Water and Sanitation encourages ongoing improvement of the management of drinking water services in South Africa through the measurement of five groups of risk indicators (Department of Water and Sanitation, 2015). The Phalaborwa, Lulekani and Namakgale assessment area water supply systems, for which Lepelle Northern Water is responsible, is operating at 101.3% of its 76 megalitres per day (Ml/d) design capacity. This is evidence of non-compliance that must be addressed by the water services authority, the Mopani DM. The water supply system achieved excellent results (99.3%) on the microbiological compliance risk indicator, but the indicators for microbiological monitoring compliance and chemical monitoring compliance were only 55% and 22.4%, respectively. The non-compliance for chemical monitoring compliance "…indicates a high-risk [sic] for the end consumers" (Department of Water and Sanitation, 2022a, p. 230). There is also non-compliance with the water safety plan risk indicator by the total absence of such a plan. It is essential that a water safety plan be developed. The technical skills risk indicator at 75% is deemed to be adequate. Notwithstanding these shortcomings, the water services authority received a Blue Drop Risk Rating of 43.7%, placing it in the low-risk category (Department of Water and Sanitation, 2022a).

As with potable water, the Mopani DM is the water services authority responsible for wastewater. The existing wastewater network is incapable of accommodating the volume of wastewater generated in the town (Ba-Phalaborwa Municipality, 2021a). The Green Drop regulation programme that seeks to improve

[4] The investigations were ongoing as of June 2022 (Special Investigating Unit, 2022).

the management of wastewater in South Africa bemoaned the critical state of the Phalaborwa wastewater treatment works (WWTW). The failings at the WWTW are not new as issues with the sanitation infrastructure in town were flagged in the 2010 IDP document (Ba-Phalaborwa Municipality, 2010) and the 2013 Green Drop report classified the plant as 'critical', which meant that it was not in compliance with legal and best-management practices, resulting in harmful effects on the receiving water resource and its downstream users (Water Group and The Association for Water and Rural Development, 2018). The causes of the non-compliance were listed as "...poor maintenance, lack of systems and monitoring, a dysfunctional SLA between Mopani District Municipality and Ba-Phalaborwa Local Municipality and underlying institutional issues'' (The Association for Water and Rural Development, 2018, p. 6). A plan to revamp the WWTW was implemented with a warning that "[f]or any improvement of the effluent produced..., a functional Service Level Agreement (SLA) has to be in place, defining in detail the responsibilities of the WSA [water services authority] and the WSP [water services provider]. If the SLA corner stone [sic] is ignored, no return on investment will [be] received on any intervention in the Wastewater business" (The Association for Water and Rural Development, 2018, p. 6). In addition, a call for the budgeting and implementation of six key actions was sounded.

Unfortunately, it appears that the call was not heeded as the 2022 Green Drop document reports that the cumulative total of four risk indicators placed the Phalaborwa WWTW at the top end of the critical risk category at 95.5%, just 4.5% remaining before the system reaches the maximum critical state. There is no information about the percentage capacity utilisation of the 8Ml/d, but the WWTW[5] microbiological compliance is at 23% and chemical compliance at 24% (Department of Water and Sanitation, 2022b). The national Department of Water and Sanitation, as the regulator, has indicated that "...wastewater systems that receive a Green Drop score below 31% are to be placed under **regulatory surveillance**, in accordance with the Water Services Act..." (Department of Water and Sanitation, 2022b, p. 14 [highlighted text in the original]), the Phalaborwa WWTW having scored 24% in 2021. Furthermore, "[t]he Regulator is concerned about the overall poor state of wastewater services at...Phalaborwa...and the consequential impact on respective water resources. It is thus required that the WSI [water services institution] submit a detailed corrective action plan within 60 days of publishing of this report. The plan must map the activities, responsible persons, timelines and expected improvements..." (Department of Water and Sanitation, 2022b, p. 53).

Effluent from the Phalaborwa WWTW that is very likely not treated to an acceptable standard is discharged into the Selati River as is effluent from the Lulekani WWTW and the Namakgale WWTW, both of which received the same directive for an action plan. The overflow of sewage from the WWTW, together with sewage-pipe leakages and bursts as well as overflows caused by pipe blockages, contaminated the watercourses in the area. A stream, 200 metres from the entrance to a

[5]The LED strategy states that the "[s] ewage works at Phalaborwa town are severely over used [sic]..." with a design capacity of 4.8 Ml/d, but an inflow of 7Ml/d (Ba-Phalaborwa Municipality, 2013, p. 13).

Fig. 3.2 Raw sewage floating in 'Shit Creek' in July 2022. (Source: Author (2022))

four-star lodge, that eventually flows into the Kruger National Park has been dubbed 'Shit Creek' by locals for the raw sewage floating in the water and the unpleasant odour emitted from it (Fig. 3.2). The effluent in this stream has been a source of serious concern for the management of the KNP (Van Huyssteen & Johnson, 2020). Other examples of sewerage problems in town are raw sewerage overflowing in the premises and buildings of a special needs school, forcing its temporary closure, and residents being forced to dig trenches for the raw sewerage to flow away from their yards (Sibuyi, 2020a; Dreyer, 2022). Some of the sanitation infrastructure dates back to the 1950s and overflows from the ageing pipe network have ill effects on the quality of life of residents and the experiences of tourists alike.

Financial support from Germany was supposedly received to upgrade the sanitation infrastructure in town (Ba-Phalaborwa Municipality, 2013). However, this funding allocation could not be verified, and the sewage problems persist. The sewer pump station in town needs to be upgraded and the asbestos sewer pipes require replacement, especially to accommodate new developments (Ba-Phalaborwa Municipality, 2022a). While Phalaborwa is experiencing major difficulties in the provision of clean water and the removal of dirty water, the provision of electricity is a key element in the continuing functioning of these systems.

3.5.2 Electricity Provision

The local municipality is responsible for the provision of electricity in Phalaborwa whereas the surrounding settlements are serviced by Eskom (Ba-Phalaborwa Municipality, 2021a). There are some 4000 electricity connections in town, ranging from residential and business connections to 11 kilovolt (kV) bulk supply to large consumers. Complications relating to electricity provision in town were already flagged in the 2009 IDP document which also mentioned that the income from rates was insufficient for the operational and maintenance costs of the electricity supply system (Ba-Phalaborwa Municipality, 2009). The electricity distribution network is in dire need of refurbishment (see Fig. 3.3) as a significant share of the 11-kV switchgear[6] and transformer equipment has been in operation for more than four decades and "[s]ome of the 11kV circuit breakers no longer meet modern safety requirements and in fact are positively dangerous to operate …" (Ba-Phalaborwa Municipality, 2022a, p. 63). The upshot is that faults occur that lead to the disruption of electricity supply to the town because of the unreliable protective equipment.

Whereas the mines receive electricity directly from Eskom, mining operations affected by the town's erratic supply of electricity negatively affects workers and small businesses in town that supply the mine cannot execute basic business operations such as the printing of invoices (Pers. com., PMC respondent, 2022; Pers. com., Foskor respondent, 2022). The Foskor office block in town is impacted significantly by the precarious electricity supply situation (Pers. com., Foskor respondent, 2022). The tourism sector is particularly vulnerable to electrical outages. A four-star lodge had to spend R24 000 on fuel for generators during a multi-day outage in portions of the town over the Easter weekend in April 2022. It was reported that tourists check out and leave Phalaborwa when accommodation establishments are without water and electricity for days on end. The concern is that affected tourists would probably never want to visit Phalaborwa again leaving the town's tourism reputation in tatters. The financial consequences the lack of service delivery have are frightening (Pers. com., Tourism Association respondent, 2022). The disruptions caused by the towns unreliable electricity supply system compound those resulting from the loadshedding that is implemented countrywide.

Phalaborwa's increased demand for electricity is set to continue, but large parts of the network cannot ensure a reliable supply of electricity as there is only a very small reserve of electricity available to the town (Ba-Phalaborwa Municipality, 2013). Urgent upgrading of the network is required, and this includes the replacement of aging and obsolete equipment, especially at two strategically important substations for the town (Ba-Phalaborwa Municipality, 2013; 2022a, b). Moreover, because the servitude lines are not maintained difficulties arise in accessing the electricity infrastructure to do emergency repairs. The municipal staff who are expected to repair infrastructure are wholly underequipped and anecdotal evidence suggests that electricians from the mines refuse to enter municipal substations because of the risks involved (Pers. com., LCBP, 2022). Sadly, there have been fatalities of municipal staff while attending to electrical repairs (Maponya, 2012; Sibuyi, 2021a).

[6] Switchgear is directly linked to the reliability of electricity supply system and comprises switches, fuses or circuit breakers that control, protect and isolate the electrical equipment.

Fig. 3.3 Degraded transformer in President Nelson Mandela Road, Phalaborwa in July 2022. (Source: Author (2022)) When this photograph was shown to a qualified electrical engineer it elicited (Pers. com., 2022) the following urgent response. "What concerns me is the obvious degradation around the insulation of the cable connections to the transformer. If that fails, then we have the risk of exposed copper cable and potential for faults and arcing. It certainly does not look normal to me. It is worth a further inspection because transformer faults can quickly lead to more devastating problems. The insulation degradation can lead to arcing which can then lead to fires or explosions. If it is transformer oil that is leaking, then the concern is the quantity and rate at which it is leaking. The oil can 'break down' electrically and cause arcing within the transformer and then can lead to an explosion in the worst-case scenario. If it is the outside coating of the transformer that is reacting, then the risk in the long term is rust and loss of structural integrity. The insulation degradation can lead to arcing which can then lead to fires or explosions. Transformer oil is supposed to be checked routinely and I don't know if this has been done for this specific piece of equipment."

The municipality reports that the electricity master plan cannot be implemented due to financial constraints (Ba-Phalaborwa Municipality, 2022a). In fact, it is unclear if any electricity master plan even exists. The problems with providing electricity are compounded by cable theft, especially during scheduled loadshedding, and vandalism of electrical infrastructure is rife despite the security personnel deployed to guard it. Sibuyi (2022a) reported that the municipality incurred losses of R500 000 due to cable theft. This despite the municipality offering a reward of

R50 000 "… for crime tips leading to the arrest and prosecution of criminals who steal the municipality electrical infrastructure *(copper cables and transformers only)*" (Ba-Phalaborwa Municipality, 2022b Furthermore, there are fundamental problems in implementing electricity maintenance plans because of financial and material constraints, delays in procurement, staff shortages, weak performance by service providers and the impact of the COVID-19 pandemic (Ba-Phalaborwa Municipality, 2020, 2021b). It is noteworthy that there was a Local Energy Forum in the past (Ba-Phalaborwa Municipality, 2021b), but it collapsed and was never resuscitated (Pers. com., senior municipal respondent). Refuse services in Phalaborwa are also constrained as overviewed next.

3.5.3 Solid Waste Management Services

Solid waste is removed regularly in town. However, in 2011 it was reported that the municipal landfill site had reached its capacity, had exceeded its lifespan and was operating illegally. Attempts to secure a new site were unsuccessful pending the outcome of a land claim for the proposed site (Ba-Phalaborwa Municipality, 2011, 2013). By 2021 it was said that solid waste was not being disposed of properly because of the unacceptable state of the landfill site. Despite this, the operating licence for the landfill site was extended to March 2024. The municipality has been issued with an approved licence to develop a new landfill site. Eighty-nine percent of the perimeter of new, fully compliant site has been fenced thanks to financial assistance from Foskor (Ba-Phalaborwa Municipality, 2021b).

The municipal integrated waste management plan that expired in June 2020 needs to ensure that a revised plan will improve the delivery of the solid waste management service. There are, however, recurring breakdowns of old waste compactor trucks and the complication of obsolete refuse removal vehicles that need to be replaced (Ba-Phalaborwa Municipality, 2021a). Currently, only two refuse removal trucks are in a good working condition (Ba-Phalaborwa Municipality, 2022a). Efforts to promote recycling to reduce the volume of waste to landfill have been launched (Ba-Phalaborwa Municipality, 2021a). The Phalaborwa Bollanoto Recycling cooperative is working to fulfil this objective (Phalaborwa Herald, 2022a). The next subsection briefly reviews the lamentable state of the roads and green spaces in Phalaborwa.

3.5.4 Roads, Parks and Open Spaces

Pothole-ridden roads and badly deteriorated pavements in town are evidence of the decay of this essential transport infrastructure (see Figs. 3.4a–c) (Ba-Phalaborwa Municipality, 2010, 2013). The municipal roads and stormwater master plan is outdated with no funds to commission a new plan. Consequently, road maintenance is

Fig. 3.4 (**a**) Potholed residential street in Phalaborwa in July 2022. (**b**) Unkempt road verges along residential street in Phalaborwa in July 2022. (**c**) Potholed parking area in Phalaborwa in July 2022. (Source: Author (2022))

limited to pothole repair within the tight constraints of the allocated budget (Ba-Phalaborwa Municipality, 2021a, b). The patching of potholes is ongoing according to a schedule, however, the task at hand is so great that roads are left in a state of disrepair (Ba-Phalaborwa Municipality, 2013, 2019b). There are roads in town that are beyond rescue by pothole repair so requiring complete resurfacing to restore them.

Streetlights in town were replaced with energy-saving devices in the 2017/18 financial year with funds from the energy-efficient, demand-side management programme residing in the national Department of Energy (Ba-Phalaborwa Municipality, 2019b). However, when the project was completed, not all of the lights worked (Dreyer & Sibuyi, 2020). Later, it was reported at the IDP public participation conducted via social media in April 2021 that streetlights in town were still not operational (Fig. 3.5).

The 2013 LED document reported that parks in Phalaborwa were well maintained (Ba-Phalaborwa Municipality, 2013). However, the issue of poor park maintenance was raised in the IDP public participation process in April 2021 (Ba-Phalaborwa Municipality, 2021b). Open spaces in town are overgrown and clearing efforts are stymied by the lack of equipment (Fig. 3.6).

The potholed streets, the uncut vegetation along road verges, the toppled streetlights and unkempt, litter-strewn open spaces are glaringly visible indicators of infrastructural neglect in Phalaborwa. Although it has been many years since the need to arrest the neglect was first recognised by the municipality, the town's infrastructure has been flagrantly left to decay. These declines are normally attributed to the financial woes of the municipality.

Fig. 3.5 Broken streetlight pole in Phalaborwa in July 2022. (Source: Author (2022))

Fig. 3.6 Unkempt, condom-strewn park space with building rubble in Phalaborwa in July 2022. (Source: Author (2022))

3.5.5 The Financial Factors

Financial constraints, irregular expenditure and a culture of non-payment exacerbate the service delivery problems in Phalaborwa. Municipal officials opined that financial instability and the high rate of staff vacancies, such as a 45% vacancy rate in 2018/19, are to blame for the Ba-Phalaborwa LM receiving poor audit outcomes and inadequate delivery of basic services to residents. The number of municipal staff has not kept pace with the significant population growth in the municipality. The history of poor audits and allegations of corruption culminated in a raid of municipal offices and an investigation of maladministration by the Directorate for Priority Crimes Investigation, colloquially known as the Hawks, in 2021 (Human Sciences Research Council, 2022).

Residents in town are billed and they pay for water services, whereas residents in the surrounding communities and villages have low rates of payment for services (Ndlozi, 2010; Pers. com., LCBP, 2022). This was corroborated by a senior municipal respondent who added that more must be done to change the non-payment attitude while being mindful of what citizens can afford. Only 39% of water consumers are billed, the remaining 61% receiving water at no cost. This in a situation where 38% of water is lost due to leakages and wastage (The Association for Water and Rural Development, 2018). Clearly, this state of affairs is financially unsustainable as the income from billed water usage does not come close to the cost of the bulk water that is supplied. In 2018 it was reported that the Ba-Phalaborwa LM owed more than R180 million to the Mopani DM for profits from the sale of water. This results in Mopani DM defaulting on its payments to Lepelle Northern Water who in turn, throttled the water supply to town by 70% because of the non-payment.

The Ba-Phalaborwa LM has signed a monthly payment agreement of R1.7m with Mopani DM (Dreyer, 2018a). However, a study concluded that "[t[he municipality stands to gain R62.3 million per annum if…billed consumption increased by 35%" (The Association for Water and Rural Development, 2018, p. 15).

The Ba-Phalaborwa LM is in a vulnerable financial position because it cannot recover money from consumers for the services it renders, a dilemma the municipality recognises (Ba-Phalaborwa Municipality, 2021a, b, 2022a, b; Auditor-General South Africa, 2022). Furthermore, in the "…review of Ba-Phalaborwa municipality…municipal officials warned that political interference destabilised the administration of the municipalit[y] and prevented innovative management but even worse caused service delivery failures" (Human Sciences Research Council, 2022, p. 92). The Human Sciences Research Council document makes recommendations how this can be addressed. Political instability is one of the reasons provided by Rangwato et al.'s (2022) investigation into the inability of the Ba-Phalaborwa LM to effectively spend its municipal infrastructure grant allocation from national treasury. Interviewees in this study echoed the problem of political considerations trumping the needs for service delivery. Political factions in the municipality pose a serious problem as they undermine efforts to deliver services (Pers. com., Senior municipal respondent, 2022).

The high costs of repairs and maintenance to aged infrastructure will persist as a major obstacle for the municipality to negotiate and a low rate of revenue collection will worsen the situation. While residents wait for the municipality to spend its financial resources more effectively, the municipality waits for residents to start paying for services – a stalemate situation (Pers. com., Retired senior municipal official, 2020). There is the view that it is the municipality that should take the lead and provide an uninterrupted supply of water for which residents must commit to paying for (Pers. com., Senior municipal official, 2022).

While the status quo persists, it has been noted that the Ba-Phalaborwa LM has not allocated any funds for fixing basic services in town: "…despite an overall budget of some R685-million, nothing has been set aside in the proposed 2022/2023 budget, nor in the projected budgets for the subsequent 2 years …" (Let's Change Ba-Phalaborwa, n.d.-a, unpaginated). Furthermore, the most recent IDP lists capital projects related to electricity, roads and stormwater, waste management and parks as "[u]nfunded projects" (Ba-Phalaborwa Municipality, 2022a, pp. 171–172). Projects that are part of the provision of basic services are not funded or do not have funds allocated to them: "[t]he Official from the Municipality indicated that this…is essentially a "wish list" of projects that were to be undertaken "should funds be available"" (Let's Change Ba-Phalaborwa, n.d.-b, unpaginated). It is conceivable that the municipality would approach the local mining sector for assistance with these projects. It is the generous contributions by the mines to municipal service delivery which is the focus of the next section.

3.6 The Support of the Mines for Municipal Service Delivery

The social and labour plans (SLPs) of the two mines, as required by the Mineral and Petroleum Resources Development Act (Act No. 28 of 2002) together with the 2018 Mining Charter, outline how Foskor and PMC assist the local municipality in honouring its service delivery mandate. The IDP is the main identifier of municipal projects to which the mines give assistance, but there are also projects funded by the mines that are not listed in the IDP and the SLPs. However, in the near future the mines will be required to consult broadly with the local communities regarding projects and not rely solely on the IDP. This stems from the impression that the Department of Mineral Resources and Energy has that municipalities are identifying projects to be funded by the mines that should actually be funded by the municipalities themselves (Pers. com., PMC respondent, 2022).

Foskor spent R30 million on three projects between September 2014 and August 2019 that relate to electricity (R10m), waste management (R7m) and the rehabilitation of streets (R13m), with a further R7m allocated as part of the 2019–2023 SLP for the development of the new landfill site (Foskor, 2019). Similarly, PMC boasts that between 2011 and 2021 they spent R120 million on roads and electricity projects in the local municipality (PMC, n.d.). In addition, the mines avail the services of their artisans to assist the municipality whenever they are required; at no charge to the municipality (Pers. com., PMC respondent, 2022; pers. com., Foskor respondent, 2022).

The rehabilitation of streets in the municipal area has also been funded by Foskor and implemented through the Expanded Public Works Programme (Foskor, 2021). Foskor by allocating a budget of R1m and PMC (in collaboration with the municipality) have established a Pothole Repair Project in town "…in a respond [sic] to the infrastructure problems prevailing in Phalaborwa…" (Pala Connector, 2016/2017, p. 9; Foskor, 2019). The Foskor respondent stated that "we are repairing potholes on a permanent basis" (Pers. com., 2022). In 2018/2019 PMC sponsored the rehabilitation of the 2.2-kilometre Park Street (in which the PMC-built MarulaMed Health Centre is located) and in 2020 the kilometre-long Molengraaf Street (feeder road to the road entering the PMC mine) (Palabora Mining Company, n.d.; PMC, n.d.). The rehabilitated Molengraaf Street was presented to the mayoress at a ceremony in January 2020 (Pala Connector, 2020). PMC also sponsored the rehabilitation of 350 metres of Potgieter Street as well as a 100-metre stretch, of the R530 (Phalaborwa Herald, 2022b).

PMC, in collaboration with the Palabora Foundation (the corporate social responsibility arm of the company), purchased, at the request of the local municipality, a jet sewer machine for unblocking drains and dealing with overflows from sewer pipes. The PMC manager for transformation, stakeholder engagement and internal communication proclaimed that it was "…vital that we maintain the partnership between the business [PMC] and the local municipality. The challenges experienced by our municipality also affect us both on a personal and a business level, therefore we continue to fight for a better standard of living for our people" (Pala Connector, 2021, p. 15). This sentiment was echoed by the PMC transformation superintendent who was "…excited at the idea of handing over the long-awaited machine: [w]hen we received the letter from the Municipality requesting…the machine, we did not hesitate to respond positively to their plea because we understood the sense of urgency…" (PMC, 2021, p. 1). The municipality was appreciative of the donation of the jet sewer machine against the backdrop of the COVID-19 pandemic and the water services manager thanked PMC "…for helping us in keeping our town clean. It is heartbreaking to see our roads flooded with water, which ultimately ruins the quality of the roads. The Jet Sewer Machine will…assist us in maintaining good hygiene and sanitation" (PMC, 2021, p. 1.). The water services manager explained that it would take some time to overcome the sanitation crisis as the municipality did not have the necessary resources to meet the challenge. The required resources included a tractor loader backhoe (a digger) and a water pump. The national Department of Forestry, Fisheries and the Environment donated the digger and refuse trucks to the Ba-Phalaborwa LM in June 2022 (Phalaborwa Herald, 2022c).

PMC has funded the upgrading and refurbishment of electrical transformers in town (Pers. com., PMC respondent, 2022). The failure at the Selati 11 kV main intake substation led to Foskor consulting with the municipality to remedy the situation at a cost of R10m and to contribute to the upgrading and refurbishing of the electricity network (Foskor, 2021; Pers. com., Foskor respondent, 2022). The company has also assisted the municipality with waste management in various ways. It has relinquished the surface right to 50 hectares on the Schiettocht farm for the establishment of a new landfill site for the local municipality. This was done with the aim of cultivating a culture of 'no more waste' and the effective management of waste from the creation point of the waste to its dumping point (Foskor, 2021). Foskor purchased a refuse truck and handed it over to the municipality in 2018 as part of the landfill site project. The second phase of the project was the construction of concrete palisade fencing along 3.5 km of the perimeter of the site in 2019 with the completion of the remaining 400 metres scheduled for 2020 (Foskor, 2019, 2020). Unfortunately, further work on the landfill site was halted because Foskor encountered financial problems, but the project is set to resume in the latter part of 2022 (Pers. com., Foskor respondent, 2022). A clean-up campaign initiated by a public-private partnership between the municipality and PMC was launched in January 2022 with the aims of revamping the town and "…to attract potential investors, business opportunities and tourism…to revive the economy…to its glory days whereby it used to create jobs" (Modiba, 2022, s.p.). The 'glory days' refer to the times when the Ba-Phalaborwa LM won the provincial and national cleanest town competition in 2005 and were placed first in the provincial awards in 2008 and second in 2006 and 2007 (Ba-Phalaborwa Municipality, 2009, 2011). At the launch clean-up campaign PMC committed its readiness to further assist the municipality with waste collection trucks and bins as part of a wider cleaning campaign to which all stakeholders would be invited (Modiba, 2022). It was reported in early July 2022 that PMC had donated 30 skips and a new refuse collection truck as part of the company's commitments to ongoing assistance to the municipality (Phalaborwa Herald, 2022a). It is unclear whether the invited stakeholders would include civil society groups in Phalaborwa.

3.7 Civil Society and Municipal Service Delivery

Phalaborwa has an eventful history of residents' associations calling for the problems with service delivery in the town to be dealt with. There have been protests, marches and petition handovers that have been covered by the media. The associations have used instant messaging services, social media platforms and websites to inform and mobilise the community. Residents have formed groups to clean the town themselves. Despite years of attempts at interaction with the municipality to halt the decline, it has been of little or no avail.

Members of the ratepayers' association (RPA) have withheld payment of their property taxes since May 2008 in protest against the poor service delivery in Phalaborwa, but they have continued to pay for electricity and other services. The municipality responded by disconnecting the electricity supply to the property tax defaulters, despite court orders to restore the service (Morkel, 2010). Ndlozi (2010) has pointed out that the RPA at the time was not recognised by the municipality. Back then, unity among the members of the RPA was fragile and support from the community was lacking. Anecdotal evidence suggests that the same difficulties were reported by the RPA in 2020. Residents would complain about the lack of service delivery (sewerage problems, potholes and unkempt parks) but expected others to do something about it. At a meeting in April 2010 it was reported that there were 1888 applications for membership of the then RPA, but only 499 (26%) of the applicants had paid the membership fee. The issue of payment of withheld rates and taxes to the RPA, rather than to the local municipality, was a discussion point at the meeting, with only 200 of the paid-up members having deposited their rates and taxes into the RPA's account. The total amount (R700 000) collected was deemed insufficient to effect any changes in the town (Albert, 2010). In 2012 the RPA complained about the irregular maintenance of underground infrastructure and requested a report on municipal infrastructure to enable the RPA to do effective oversight of municipal maintenance work and the costs involved (Phalaborwa Herald, 2012). It is not known whether the municipal infrastructure report was ever provided to the RPA.

In 2015 the Phalaborwa Homeowners' Association (PHOA) launched a campaign challenging the validity of the property valuation roll. Together with the RPA, they organised marches of protest over the failing infrastructure and the 641% increase in property taxes over 3 years (Kelder, 2015; Dreyer, 2017, 2018b). On the occasion of one of the marches the then mayor, Pule Shayi, declared that the municipal council had to work in partnership with residents to ensure that the proper municipal services are rendered and that the residents should hold the council accountable (Dreyer, 2017).

In 2018 the RPA received 330 letters from primary school learners in which they appealed to the mayor for water to wash themselves and electricity to help them do their homework at night. The letters were affixed to cardboard posters that were displayed along the perimeter fence and in the parking lot of the municipal offices. As the mayor stood on the platform to address the crowd, he "…fell silent. Looking visibly emotional…[he] requested to approach one of the posters to read the letters. In silence he walked off the platform and approached the board. He stood there staring at one particular letter, scribbled in blue crayon over a white sheet; the letters read 'My Mayor, plz, water'. A heartfelt plea to the mayor – from his very own fourth grade daughter" (Dreyer, 2018c, p. 3). A petition from the RPA, on behalf of the residents and learners of Phalaborwa, calling for an investigation into the provision of water and other municipal services was tabled in the National Assembly on 8 November 2018, but it was referred to the Portfolio Committee on Water and Sanitation and Portfolio Committee on Cooperative Governance and Traditional Affairs for consideration and report (Parliament of the Republic of South Africa, 2018). The RPA took their service delivery grievances all the way to parliament, but nothing has happened.

In addition to the RPA and the PHOA, a residents' group calling themselves by the Trumpesque-sounding "Let's Make Phalaborwa Great Again" was formed in March 2017 to keep the town clean. The group had been organising weekly clean-up events after seeking permission from the municipality to do so. A local business donated decorative stones to be placed at the entrance of town. Initially, municipal workers provided assistance with the clean-ups but stopped after four events. Unfortunately, the municipality could not supply refuse bags as it only receives 2000 bags for the entire municipality for the year. The new group, who relied on donations of refuse bags and water, employed a permanent worker and called upon shops in town to clean their own pavements (Sibuyi, 2018). Without any assistance from the municipality the group battled financially to keep a service afloat that is a function of the municipality (Sibuyi, 2020b).

Another citizen-led initiative was launched by a tourism student in Phalaborwa who, together with 12 other students, repaired potholes in town after witnessing an accident caused by the potholes. Bags of tarmac for the pothole repairs were donated by a local businessperson (SABC News, 2020). It is clear that citizens, citizen organisations and businesses in Phalaborwa have, over the years, attempted to remedy the deterioration of urban infrastructure resulting from the local municipality's dereliction of duty.

Phalaborwa was without electricity for 5 days over the festive season in 2020. Residents lost thousands of Rands as food spoilt in fridges and freezers that were not working. Home security systems were comprised as back-up battery systems failed. Citizen anger was vented in comments posted on social media platforms. In 2020 the RPA had a database with over 3000 complaints about service delivery that was "…delivered to the Council and nothing came of it except empty promises" (Pers. com., past chairperson of the RPA, 2022). The infrastructure failures continued despite many meetings, protests, marches and petitions over the years by the RPA and other organisations, notwithstanding citizens' and businesses' repair and clean-up interventions. In February 2021 the RPA petitioned the municipality for the removal of the then municipal manager, Maite Moakamela (Sibuyi, 2021b). The Hawks raided the municipal offices on 26 February 2021 on allegations of tender corruption (Sibuyi, 2021c). The RPA disbanded in June 2021. The municipal manager resigned from her position effective 31 December 2021 and an acting municipal manager, Dr. Kgoshi Lucas Pilusa, was appointed (Modiba, 2021). The post of municipal manager was advertised in January 2022 and Dr. Pilusa was appointed for a five-year term on 1 September 2022 (Sibuyi, 2022b).

A new community group focussing specifically on Phalaborwa, Let's Change Ba-Phalaborwa (LCBP), was launched in March 2022 (Phalaborwa Herald, 2022d). The catalysts for the formation of LCBP were the decay and collapse of infrastructure and service delivery in Phalaborwa and its central business district (CBD); the indifferent response from the municipality regarding the state of the town; and the community's anger and frustration (Let's Change Ba-Phalaborwa, n.d.-b). LCBP is a non-political, non-profit forum, registered as a non-profit company committed to providing communication to the community and the municipal Council on issues that affect the community of Phalaborwa. The aim is to develop a collaborative, non-confrontational working relationship to find and implement solutions to the many persistent questions concerning the town (Let's Change Ba-Phalaborwa, n.d.-b). The core values of the forum are inclusivity (the entire community working together), productivity (using skills and experience in the community for change), sustainability (a process of engagement), community development (creating opportunities and new ways of doing things) and communication (providing information) (Let's Change Ba-Phalaborwa, 2022a).

The strategy of LCBP is rests on eight key pillars. The first pillar is 'service delivery failures' being reported via ward-based WhatsApp groups and logged on a database for action and feedback. The data from the 'incident register' is used to engage with the municipality about where the infrastructure problems are occurring. Two thirds of the service delivery incidents reported to the LCBP in the 92 days from 1 May 2022 to 31 July 2022 were related to water and sewerage (Fig. 3.7), and roads, parks and open space issues accounted for 22%, electricity-related issues for nine per cent and solid waste for only one per cent of the reports in the incident register (Let's Change Ba-Phalaborwa, 2022b). A WhatsApp group has been established to communicate with the acting municipal manager of the Ba-Phalaborwa LM and the municipal manager of Mopani DM on critical and ongoing infrastructure issues. LCBP has also established a relationship with Lepelle Northern Water.

The second pillar, 'participatory governance', refers to the right of citizens to participate in the affairs of local government as provided for in the Municipal Systems Act (Act 32 of 2000), particularly engagement in the IDP and the municipal budget processes (Let's Change Ba-Phalaborwa, n.d.-b). LCBP reports that it has managed to influence the IDP of the Ba-Phalaborwa LM, but not its budgetary component. While LCBP voices its displeasure with the state and governance of the town, it also extends a hand of cooperation to the Ba-Phalaborwa LM by asking the municipality to consider the establishment of an advisory committee of stakeholders and experts (engineering, accounting, etc.) from businesses and the community. Although the mayoress expressed a willingness to accept the offer, the municipality is yet to respond, the latter's elected councillors being described by the LCBP as "…reluctant participants" (Pers. com., LCBP respondents, 2022). Despite the mayoress's readiness to help, LCBP expressed concern: "It would be remiss not to address the issue of the attitude and behaviour of the Mayor – the political head of

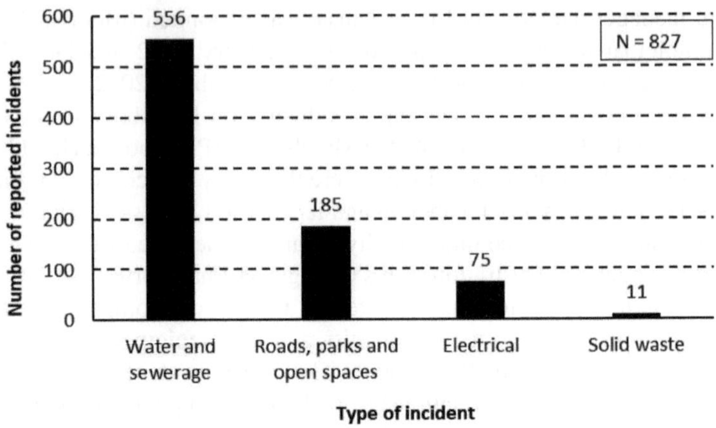

Fig. 3.7 Citizen-reported service delivery issues in Phalaborwa, 1 May 2022 to 31 July 2022. (Source: Let's Change Ba-Phalaborwa (2022b))

the municipality. You as the Phalaborwa community must be informed that the mayor continues to treat the Phalaborwa community with flippant disdain and disrespect" (Let's Change Ba-Phalaborwa, 2022c, p. 3).

The third strategic pillar is 'stakeholder engagement' where the LCBP steering committee has introduced themselves to Foskor and PMC managers and appealed to them to consider funding more projects in the town. This engagement also calls for mine employees to encourage their employers to spend some of the companies' corporate social investment funds to restore the town. LCBP wants the mining companies to be part of the proposed advisory committee for making recommendations to the Ba-Phalaborwa LM (Pers. com., LCBP, 2022). The fourth pillar, 'provincial and national interventions', relates to water, sanitation and electricity. As example, LCBP has approached the National Department of Water and Sanitation and the National Energy Regulator of South Africa (NERSA). While the former is yet to reply, NERSA met with LCBP and Ba-Phalaborwa LM representatives in a joint meeting in July 2022 (Let's Change Ba-Phalaborwa, 2022c). The meeting concluded with NERSA requesting a plan of action from the municipality regarding the resolution of electricity issues in town and the establishment of an end-users' group with collaboration between the Ba-Phalaborwa LM and LCBP on the implementation and management of the plan. The end-users' group would meet every two months and NERSA has affirmed that non-compliance would result in the revocation of the municipality's electricity distribution licence.

The fifth strategic pillar is 'service delivery options'. LCBP contends that the Ba-Phalaborwa LM can define and establish an internal municipal service district (IMSD) for Phalaborwa, as per section 85 of the Municipal Systems Act, because the LCBP sees the priorities of the town to be different to those in the rest of the local municipality. An IMSD would give the community a greater say in and control over municipal governance and service delivery in the town. Burning questions on this matter would be: Would an IMSD create inequality and inequity between town,

Namakgale, Lulekani and the traditional authority areas? Would the town flourish while the outlying areas lag behind? Does the IMSD contribute to spatial justice? Another option for service delivery proposed by the LCBP is to put the water and electricity services out on a concession where a private company would replace, augment, maintain and operate the services for a period of thirty years, for example, during which the private company would recoup their capital expenditure and generate a profit. The infrastructure would be returned to the municipality at the end of the concession period.

Strategic pillar six is 'rule of law – court action' whereby LCBP believes that the Ba-Phalaborwa LM is operating unlawfully by not providing services to the community in a sustainable manner. LCBP is gathering information in support of any future court action. In South Africa, local municipalities have been taken to the courts by civil society organisations because of the lack of service delivery with the court's ruling in favour of the civil society organisations (Ellis, 2022; Machanik, 2022). Some small towns' business forums have bypassed their local municipalities and now deal directly with Eskom for the delivery of electricity to businesses. But this route incurs expensive legal fees and takes years to materialise (Creamer Media's Engineering News, 2019; Venter, 2021). In the Free State province, a local municipality has been placed under compulsory administration following legal action by the business forum (Khumalo, 2022). LCBP asserted that good case law is being developed by increasing numbers of South African communities taking municipalities to court for their failures in service delivery (Pers. com., LCBP, 2022).

The seventh pillar is 'active citizenry' which requires a critical mass of residents to become active citizens to stem the decline of the town because of what it terms "...bad, corrupted, and irresponsible governance..." (Let's Change Ba-Phalaborwa, 2022c, p. 4). The final pillar called 'DIY', is the multiplicity of approaches used by LCBP in the hope that the community will be encouraged to become active citizens and do things themselves, like fixing potholes and raising funds (Let's Change Ba-Phalaborwa, 2022a, b).

LCBP has mobilised the community to make known to the municipality their dissatisfaction with poor service delivery. For example, water restrictions were placed on the town on 10 May 2022 because the district municipality did not pay Lepelle Northern Water. LCBP sent an email to the Premier, the Member of the Executive Committee for Co-operative Governance, Human Settlements and Traditional Affairs, and the Minister of Water and Sanitation requesting advice on and intervention in the matter. Residents were angry. LCBP organised a march to the municipal offices on 18 May 2022 to protest against the ongoing water crisis (Shipalana, 2022). Once the outstanding debt was settled, water was restored to the town the evening on the day of the march. As with the wastewater issue mentioned earlier, the current SLA between the water services authority (the district municipality) and the water services provider (the local municipality) must be revised to provide clear direction on the responsibilities of the two entities. A possible solution to the lack of intergovernmental cooperation is to have the Ba-Phalaborwa LM as both the water services authority and the water services provider that will report to citizens at one end and directly to Lepelle Northern Water at the other, and deal with the infrastructure (Pers. com., Senior municipal respondent, 2022).

In the case of water provision, the Ba-Phalaborwa LM seems to want to be released from the shackles of an unworkable SLA. The inevitable question is whether the local municipality would be prepared to join hands with civil society forums such as the LCBP. Phalaborwa has enormous human capital which can be tapped into by the municipality to improve service delivery, guarantee cost-effective management and direct future-oriented planning: "…we sit with people who have incredible resources and skills. The knowledge base that's in Phalaborwa is a little bit ridiculous, but it's getting everyone around the same table…going in the same direction, singing from the same hymn book; what does that look like? How does it start? Where do we get it going?" (Pers. com., Tourism Association respondent, 2022). The answers to these pertinent questions and a better future hopefully lie in greater citizen involvement in the ailings and failings of the town through citizen actions and by holding local government accountable while simultaneously having civil society and local government working together.

3.8 Concluding Thoughts

The dilapidated state of Phalaborwa's infrastructure and the municipality's problems with service delivery are by no means unique. They mirror similar situations of municipal non-fulfilment of obligations in many small towns across South Africa. Consequently, there are numerous examples in the country where the affected residents and their residents' organisations have reacted by willingly assisting their local municipalities in maintaining and restoring infrastructure. However, not everyone finds this idea palatable as there is the concern that residents' groups would morph into independent political parties, in addition to the view that residents' associations eventually lose steam and fold, as happened in the past in Phalaborwa (Pers. com., Senior municipal respondent, 2022). Nevertheless, citizens are willing to step in where local municipalities are deemed to be failing.

In Phalaborwa the mining companies have been instrumental in keeping an ailing infrastructure hobbling along by assisting the local municipality to fix outages, breakages and leakages. Respondents in surveys are unanimous in their assessment that sustainable plans for various service delivery sectors must be drafted, followed by concerted implementation backed by the necessary political will to effect meaningful infrastructural turnaround. Members of residents' groups have been volunteering their time, labour and own finances, and businesses have provided financial and material assistance. Although this model is unsustainable and unfair to citizens, communities have the constitutional right to participate in governance and to hold municipalities to account for the non-execution of their mandates.

The Let's Change Ba-Phalaborwa (LCBP) forum is exploring various avenues to hold the municipal administration accountable for their mismanagement of Phalaborwa and the municipality while simultaneously extending the offer of expertise to partner with the municipality. This scenario throws up a number of questions for consideration. Would the municipality be amenable to the accountability-partnership approach? Would the fractious relationship between the residents' associations, municipal administration and the local politicians continue to stymie attempts to turn the town around? What would the best way be to get the local and district municipal administrations and the respective politicians, traditional authorities, the mining companies and residents' associations such as the LCBP to work together for a brighter future for Phalaborwa and the Ba-Phalaborwa Local Municipality as a collective? How can the infrastructural problems in town be addressed together with the myriad challenges in the rest of the municipality, all with spatial justice in mind? How can there be assurance that the needs of the community would be placed ahead of political considerations? Can citizens' collectives learn from approaches followed by the residents' groups in other towns across the country to force municipalities to take responsibility for their service delivery failures? Does the country need guidelines on 'How to hold local municipalities accountable' and 'How local municipalities can work together with citizens' groups'?

The dedicated maintenance of infrastructure, the clearance of infrastructure backlogs and the renewal of ageing infrastructure, coupled to the resolution of financial issues of the municipality, must be addressed if there is any hope of Phalaborwa being restored to its rightful place as a provincial development node. The challenges will not be overcome overnight, but at the very least the local municipality can embark on a programme of repair and rehabilitation guided by proper plans, sufficient budgetary allocation and an ingrained culture of service excellence. This is the way for the community's faith in the municipality to be restored. The LCBP forum captures this by maintaining that they "…are very mindful that unless we can return the municipality to good, clean, responsible, pragmatic, and progressive governance, we will never bring an end to our service delivery and infrastructural challenges…" (Let's Change Ba-Phalaborwa, 2022c, p. 2). A Phalaborwa with a functioning infrastructure, possibly brought about by greater collaboration between civil society and the local municipality, would be the catalyst to unlocking the future economic growth of the town and its surrounds.

References

Albert, M. (2010). Gemoedere loop hoog op vergadering van BBV [Emotions run high at RPA meeting]. *Phalaborwa Herald*, 2 April, p. 3.

Auditor-General South Africa. (2022). *Consolidated general report on local government audit outcomes: MFMA 2020–21*. Auditor-General. Accessed online June 25, 2022. https://www.agsa.co.za/Portals/0/Reports/MFMA/2020-21/FINAL_MFMA%202020-21%20GR_15%20June_2022%20tabling.pdf?ver=2022-06-15-095648-557

Ba-Phalaborwa Municipality. (2009). *Approved 2009/10 IDP*. Ba-Phalaborwa Local Municipality. Accessed online July 26, 2022. https://web.archive.org/web/20100525090529/http://www.ba-phalaborwa.gov.za/docs/IDP/APPROVED_IDP_2009-10.pdf

Ba-Phalaborwa Municipality. (2010). *Approved IDP document 2010–2015*. Ba-Phalaborwa local Municipality. Accessed online July 14, 2022. http://mfma.treasury.gov.za/Documents/01.%20 Integrated%20Development%20Plans/2010-11/02.%20Local%20Municipalities/LIM334%20 Ba-Phalaborwa/LIM334%20Ba-Phalaborwa%20-%20IDP%20-%201011.pdf

Ba-Phalaborwa Municipality. (2011). *Approved IDP document 2011–2016*. Ba-Phalaborwa local Municipality. Accessed online July 14, 2022. https://www.yumpu.com/en/document/read/14009987/approved-idp-doc-2011-16ba-phalaborwapdf

Ba-Phalaborwa Municipality. (2013). *Ba-Phalaborwa local economic development strategy*. Ba-Phalaborwa Local Municipality. Accessed online January 17, 2022. http://www.phalaborwa.gov.za/docs/tariffs/LED%20STRATEGY%20REVIEW%20Complete%20doc.pdf

Ba-Phalaborwa Municipality. (2019a). *Municipal spatial development framework 2019–2024*. Ba-Phalaborwa Local Municipality. Accessed online August 3, 2022. http://www.phalaborwa.gov.za/docs/SDF/BPLM%20SDF%20FINAL%20DRAFT%2030TH%2007%202019%20 print2%20(1).pdf

Ba-Phalaborwa Municipality. (2019b). *2017/18 Annual report*. Ba-Phalaborwa Local Municipality. Accessed online January 17, 2022. http://www.phalaborwa.gov.za/docs/reports/FINAL%20 %202017-18%20%20ANNUAL%20REPORT%2006%20APRIL%202019.pdf

Ba-Phalaborwa Municipality. (2020). *2018/19 Draft annual report*. Ba-Phalaborwa Local Municipality: Phalaborwa. Accessed online January 17, 2022. http://www.phalaborwa.gov.za/docs/reports/2018-19%20DRAFT%20ANNUAL%20REPORT%20-%2021%20JAN%20 2020.pdf

Ba-Phalaborwa Municipality. (2021a). *2019/20 Draft annual report*. Ba-Phalaborwa Local Municipality. Accessed online June 14, 2022. http://www.phalaborwa.gov.za/docs/reports/DRAFT%20ANNUAL%20REPORT%202019%202020210610.pdf

Ba-Phalaborwa Municipality. (2021b). *2020/21 Annual report*. Ba-Phalaborwa Local Municipality. Accessed online June 26, 2022 http://www.phalaborwa.gov.za/docs/reports/2020%2021%20 ANNUAL%20REPORT0323.pdf

Ba-Phalaborwa Municipality. (2022a). *IDP 2022/23–2027*. Ba-Phalaborwa Local Municipality. Accessed online June 26, 2022. http://www.phalaborwa.gov.za/docs/idp/FINAL%20 2022-23%20IDP-26%20MAY%202022_060622.pdf

Ba-Phalaborwa Municipality. (2022b). *Reward for reporting theft of municipal electricity infrastructure (copper cables and transformers)*. Ba-Phalaborwa Local Municipality. Accessed online January 6, 2023. https://www.phalaborwa.gov.za/docs/notices/20220803101647102.pdf

Beauclerk, J., Pratt, B., & Judge, R. (2011). *Civil society in action: Global case studies in a practice-based framework*. INTRAC.

Bird, K. (2021, January 22). Adopt a pothole in Secunda. *Ridge Times*.

Boraine, A. (2014). *What's gone wrong? South Africa on the brink of failed statehood*. New York.

Bulbulia, T. (2020, November 3). Rainbow, Bosveld to co-develop South African rare earths project. *Creamer Media's Mining Weekly*.

Cartwright, A. P. (1972). *Phalaborwa: Mining city of the future*. Purnell.

Cartwright, A. P. (1986). *Phalaborwa: A mining success story*. Phalaborwa.

Creamer Media's Engineering News. (2019, October 28). Eskom and a group of business strike a landmark deal, sidelining the local municipality. *Creamer Media's Engineering News*.

Delius, P. (1984). *The land belongs to us: The Pedi polity, the Boers, and the British in the nineteenth-century Transvaal*. University of California Press.

Delius, P., Maggs, T., & Schoeman, A. (2014). *Forgotten world: The stone-walled settlements of the Mpumalanga escarpment*. Wits University Press.

Department of Water and Sanitation. (2015). *Factsheet: The Blue Drop: Highlights and trends from 2009 to 2014*. Department of Water and Sanitation. Accessed online July 3, 2022. https://www.wrc.org.za/wp-content/uploads/mdocs/The%20Blue%20Drop%20Factsheet.pdf

Department of Water and Sanitation. (2022a). *Blue Drop: Progress report 2022*. Department of Water and Sanitation. Accessed online April 20, 2022 https://ws.dws.gov.za/IRIS/releases/2021_BD_PAT_report_final-28Mar22_MN_web.pdf

Department of Water and Sanitation. (2022b). *Green Drop 2022: Limpopo*. Department of Water and Sanitation. Accessed online June 23, 2022 https://ws.dws.gov.za/IRIS/releases/GD22%20Report_Limpopo_Rev03_05May22_MN%20web.pdf

Dreyer, J. (2017, November 10). Will Shayi be Phalaborwa's saviour? *Far North Bulletin*.

Dreyer, J. (2018a, October 12). 'We have reached a point beyond gatvol'. *Far North Bulletin*.

Dreyer, J. (2018b). Water crisis update: BPM owes Mopani R180m. *Far North Bulletin*.

Dreyer, J. (2018c, October 28). Mayor's own daughter pleads for water. *Far North Bulletin*.

Dreyer, J. (2022, January 14). Same sewerage different year. *Far North Bulletin*.

Dreyer, J. & Sibuyi, B. (2020, October 16). "To install lights, not make them work". *Far North Bulletin*.

Drummond, J., & Nel, V. (2021). Mafikeng: Where traditional leadership and development frameworks collide. In A. Lemon, R. Donaldson, & G. Visser (Eds.), *South African urban change three decades after apartheid* (pp. 197–214). Springer.

Ellis, E. (2022, March 14). Judge to oversee the financial recovery of 'failed' Makana Municipality. *Daily Maverick*, https://www.dailymaverick.co.za/article/2022-03-14-judge-to-oversee-the-financial-recovery-of-failed-makana-municipality/

Evers, T. M., & Van der Merwe, N. J. (1987). Iron Age ceramics from Phalaborwa North Eastern Transvaal Lowveld, South Africa. *The South African Archaeological Bulletin, 42*(146), 87–106.

Foskor. (2019). *2019 integrated report for the year ended 31 March*. Accessed online June 28, 2022. http://www.foskor.co.za/Annual%20Reports/Annual%20Report%202019.pdf

Foskor. (2020). *Integrated report 2020 for the year ended 31 March*. Accessed online June 28, 2022. http://www.foskor.co.za/Annual%20Reports/Integrated%20Report%202020.pdf

Foskor. (2021). *Economic empowerment*. Accessed online June 28, 2022. http://www.foskor.co.za/SitePages/Sustainability/Economic%20Empowerment.aspx

Grewar, J. F. (1980). *The vanished saddleback: The story of Phalaborwa*. Purnell.

Human Sciences Research Council. (2022). *Municipal skills and capacity assessment study*. HSRC. Accessed online April 5, 2022. https://repository.hsrc.ac.za/bitstream/handle/20.500.11910/19271/12231%281%29.pdf?sequence=1&isAllowed=y

Karrim, A. (2021, November 2). The angry ones: Volunteer group-turned-independent candidates fix Harrismith themselves. *News24*.

Kelder, K. (2015, November 2015). Phala residents take municipality to court. *Letaba Herald*.

Khumalo, J. (2022, April 29). Court compels Free State premier to place ailing Mafube municipality under administration. *News24*.

Killick, D., Miller, D., Thondhlana, T. P., & Martinón-Torres, M. (2016). Copper mining and smelting technology in the northern Lowveld, South Africa, ca. 1000 CE to ca. 1880 CE. *Journal of Archaeological Science, 75*, 10–26.

Lepelle Northern Water. (2021). *Annual Report 2020–2021*. Lepelle Northern Water. Accessed online June 23, 2022. https://static.pmg.org.za/Annual_Report_LNW_2021_Final.pdf Accessed 23 June 2022.

Let's Change Ba-Phalaborwa. (2022a). *Let's Change Ba-Phalaborwa*. Accessed online May 24, 2022. http://www.lcbp.co.za/

Let's Change Ba-Phalaborwa. (2022b). *Incident register: Service delivery issues*. Let's Change Ba-Phalaborwa.

Let's Change Ba-Phalaborwa. (2022c). *Newsletter update*. 12 July. Accessed online July 17, 2022. http://www.lcbp.co.za/sitepad-data/uploads//2022/07/LCBP-Newsletter.pdf

Let's Change Ba-Phalaborwa. (n.d.-a). *Open letter to the residents of Wards 11 & 12 of Phalaborwa, and the Speaker and Mayor of Ba-Phalaborwa Municipality*. Phalaborwa.

Let's Change Ba-Phalaborwa. (n.d.-b). *Interim submission by "Let's Change Ba-Phalaborwa" forum in response to the draft 2022/2023 IDP Projects presentation published by Ba-Phalaborwa Municipal Council*. LCBP.

Lindeque, B. (2021). Senekal community rally together – restoring hearts by restoring their town! *Good Things Guy*. Accessed online July 13, 2022. https://www.goodthingsguy.com/environment/senekal-community-is-rally-together-and-restoring-hearts-by-restoring-their-town/

Machanik, P. (2022, June 23). Make miscreant leaders pay for broken municipalities. *Mail & Guardian.*

Maponya, F. (2012, November 14). Municipality's electrical unit head electrocuted. *Sowetan Live.*

Miller, D. (1995). 2000 years of indigenous mining and metallurgy in southern Africa. *South African Journal of Geology, 98*(2), 232–238.

Modiba, M. (2021, December 2021). Ba-Phalaborwa municipal manager, Ms. Maite Moakamela resigns. *Ba-Phalaborwa News.*

Modiba, M. (2022, January 17). Palabora Mining Company (PMC) and Ba-Phalaborwa Municipality launched a cleaning campaign. *Ba-Phalaborwa News.*

Moffett, A. J. (2017). *Phalaborwa where the hammer is heard': Crafting together the political economy of Iron Age communities in southern Africa, AD 900–1900* [Doctoral dissertation]. University of Cape Town.

Moffett, A. J., Hall, S., & Chirikure, S. (2020). Crafting power: New perspectives on the political economy of southern Africa, AD 900–1300. *Journal of Anthropological Archaeology, 59*(101), 180.

More, C. E. (1974). Some observations on 'ancient' mining at Phalaborwa. *Journal of South African Institute of Mining and Metallurgy, 74*(6), 227–232.

Morkel, A. (2010, August 26). *Kragafsnydings as hefboom gebruik* [Power cuts used as leverage]. Afrikaner.

Moroaswi, Q. M. (2013). *A financial analysis of management models in land reform in South Africa.* [Masters thesis]. University of Pretoria.

National Treasury. (2022). *Briefing by NT on underspending on municipal revenue management improvement programme (MRMIP) as at the end of the 2021/22 financial year, 21 September.* Accessed online January 2, 2023.

Ndlozi, N. D. (2010). *Service delivery enhancement: A case study of Ba-Phalaborwa Municipality in Mopani District of Limpopo* [Masters thesis]. University of Limpopo.

Pala Connector. (2016/2017). Local economic development initiatives. *Pala Connector*, December/January. Accessed online June 14, 2022. https://www.palabora.com/documents/Palabora-Connector-Newsletter-Dec16-Jan17.pdf

Pala Connector. (2020). PMC roads rehabilitation: Service delivery at heart. *Pala Connector*, January/March. Accessed online June 14, 2022. https://www.palabora.com/documents/Pala-Connector-Newsletter-January-March-2020-edition_Covid-19-Edition.pdf

Pala Connector. (2021). A new Jet Sewer Machine aims to end sanitation challenges at Ba-Phalaborwa Municipality. *Pala Connector*, March. Accessed online June 14, 2022. https://www.palabora.com/documents/Pala%20Connector%20Newsletter%20-%2031%20MARCH%202021%20EDITION_online%20Broadcast.pdf

Palabora Mining Company. (n.d.). *Social and labour for Palabora Copper Pty (Ltd) 2018–2022.* PMC: Phalaborwa. Accessed online June 14, 2022. https://www.palabora.com/documents/PC%20SLP%202018%20-%202022_.pdf

Parliament of the Republic of South Africa. (2018). Announcements, tablings and committee reports, Thursday, 8 November. Parliament of the Republic of South Africa: Cape Town. Accessed online July 12, 2022. https://www.parliament.gov.za/storage/app/media/Docs/atc/4586d8dd-1f60-49bb-99e8-851de37cab46.pdf

Phalaborwa Herald. (2012, January 27). Shocking service delivery marks start of new year. *Phalaborwa Herald.*

Phalaborwa Herald. (2022a, July 1). Palabora Copper Mining donates skips and trucks. *Phalaborwa Herald.*

Phalaborwa Herald. (2022b, April 1). Potgieter Street in Phalaborwa town is complete and handed to municipality. *Phalaborwa Herald.*

Phalaborwa Herald. (2022c, June 10). Municipality receives TLB and refuse trucks. *Phalaborwa Herald.*

Phalaborwa Herald. (2022d, March 25). A call for residents to join new initiative. *Phalaborwa Herald.*

Plug, I. (1996). Changing societies in South Africa during the 18th, 19th and early 20th centuries: A faunal perspective. *South African Journal of Ethnology, 19*(2), 95–100.

PMC. (2021). *A Jet Sewer Machine that aims to end sanitation challenges in Ba-Phalaborwa Municipality*. Accessed online June 14, 2022. https://www.palabora.com/documents/Press%20Release%20A%20Jet%20Sewer%20Machine%20that%20aims%20to%20end%20sanitation%20challenges%20in%20BaPhalaborwa%20Municipality%2015%20March%202021%20final.pdf

PMC. (n.d.). Social and labour plans. Accessed online June 14, 2022. https://www.palabora.com/social_and_labour_plan.asp

Pottinger, B. (1990). *Loolekop: The story of Palabora Mining Company*. Jonathan Ball Publishers.

Rangwato, T. E., Mukonza, R., & Molepo, J. (2022). The spending of municipal infrastructure grants at Ba-Phalaborwa municipality. *Journal of Local Government Research and Innovation, 30*, a88.

Republic of South Africa. (1970). *Declaration of a national monument*. Government Gazette, No. 2928, Notice no. 2014 and 2015, 20 November, p. 12. Government Printer: Pretoria.

Republic of South Africa. (2003). *Local government: Municipal finance management act (Act 56 of 2003)*. Government Gazette, No. 26019, Notice no. 176, 13 February 2004. Government Printer: Pretoria.

Roux, E. H., De Jager, D. H., Du Plooy, J. H., Nicotra, A., Van der Linde, G. J., & De Waal, P. (1989). Phosphate in South Africa. *Journal of the Southern African Institute of Mining and Metallurgy, 89*(5), 129–139.

SABC News. (2020). *A group of students in Phalaborwa use their time to fix potholes*. 3 December. Accessed online June 20, 2022. https://www.youtube.com/watch?v=vm620N14rjU

SANews. (2022, September 21). 43 South African municipalities have collapsed, 151 on the brink – Treasury. *SANews*.

Scully, R. (1978). *Phalaborwa oral tradition*. [Doctoral dissertation]. State University of New York at Binghamton.

Shipalana, T. (2022, May 20). No water for over a week. *Phalaborwa Herald*.

Sibuyi, B. (2018, February 9). Operation town clean-up: 'We are fed up'. *Letaba Herald*.

Sibuyi, B. (2020a, December 11). Disabled children suffer under Phalaborwa housing and health departments. *Far North Bulletin*.

Sibuyi, B. (2020b, October 19). BPM must pay for private services. *Far North Bulletin*.

Sibuyi, B. (2021a, February 7). Phalaborwa electrician electrocuted at Selati sub-station. *Far North Bulletin*.

Sibuyi, B. (2021b, February 5). Citizens rise up against Moakamela. *Far North Bulletin*.

Sibuyi, B. (2021c, February 26). Hawks lock down BPM offices. *Far North Bulletin*.

Sibuyi, B. (2022a, October 2). Newly appointed Pilusa says heads will roll. *Far North Bulletin*.

Sibuyi, B. (2022b, September 5). New MM for BPM. *Far North Bulletin*.

Simelane, H. Y., & Sihlongonyane, M. F. (2021). A comparative analysis of the influence of traditional authority in urban development in South Africa and Eswatini. *African Studies, 80*(2), 153–171.

Special Investigating Unit (2022). *Presentation on the status of the SIU's investigations relating to the Department of Water and Sanitation, 8 June*. Accessed online August 3, 2022. https://static.pmg.org.za/220608Final_SIU_Presentation_-_Dept_of_Water_and_Sanitation_to_SCOPA_8_June_2022_1.pdf

Taylor, J. (2013). Struggles gainst systems that impoverish: South African civil society at the crossroads. *Development in Practice, 23*(5–6), 617–630.

The Association for Water and Rural Development (2018). *Capacity development for water demand & water conservation management*. The Association for Water and Rural Development: Hoedspruit. Accessed online July 14, 2022. http://award.org.za/wp/wp-content/uploads/2021/04/AWARD-TECH-REPORT-14-Capacity-Development-for-water-demand-water-conservation-management-Ba-Phalaborwa-Municipality-2018v1.pdf

The Post. (2018, May 11). Maak dorpsingange mooi [Beautifying town entrances]. *The Post*.

Van der Merwe, N. J., & Scully, R. T. (1971). The Phalaborwa story: Archaeological and enthnographic investigation of a South African Iron Age group. *World Archaeology, 3*(2), 178–196.
Van Huyssteen, C. W., & Johnson, T. L. (2020). Evaluating the use of iron-coated tubes for wetland delineation in South Africa: A pilot study in the Kruger National Park. *WaterSA, 46*(3), 393–403.
Venter, T. (2021, May 27). What are the remedies for collapsing local governments across South Africa? *Farmer's Weekly*.
Venter, T. (2022, July 9). The role of civil society in rebuilding South Africa. *Farmer's Weekly*.
Ward, C. D. (2012). *Livelihoods and natural resource use along the rural-urban continuum*. [Doctoral dissertation]. Rhodes University.
Water Group and The Association for Water and Rural Development (2018). *A turnaround plan for municipal waste water treatment: Ba-Phalaborwa Local Municipality. Project summary*. The Association for Water and Rural Development: Hoedspruit. Accessed online July 6, 2022. http://award.org.za/wp/wp-content/uploads/2020/05/AWARD-BROCHURE-WATER-GROUP-A-Turnaround-plan-for-waste-water-treatment-works-2018-v2.pdf
Weideman, M. (2015). *The changing status of civil society organisations in South Africa, 1994 to 2014*. Hans Siedel Foundation.

Chapter 4
Collaborative Governance, Social Capital and Drought: A Case Study of a Collaborative Governance Regime in Graaff-Reinet

Rebecca Light

4.1 Introduction

Rapid anthropogenic climate change has led to a global increase in extreme weather events, and with it, brought an increase in weather-related disasters (Faling et al., 2012; Aldrich & Meyer, 2015; Busayo & Kalumba, 2020). During the last decade, drought has been a disaster particularly relevant in South Africa, affecting national water security in the Western and Eastern Cape. Water security pressures are worsened by the growing population, improper planning, poor management of water resources and a lack of sufficient investment in water reservoir infrastructure (Mahlalela et al., 2020). Traditionally, policy and planning preparedness for such disaster events have been primarily focused on physical resources and infrastructure (Aldrich & Meyer, 2015). Though this is important, disaster preparedness and response are multi-faceted, and the success thereof depends on more than physical infrastructure alone. That is, solving disaster-related issues without including less-tangible capital: human and social capital is less likely to be effective (Bendimerad, 2003; Chamlee-Wright & Storr, 2011; Satterthwaite, 2011; Cretney, 2016). Social capital refers, in short, to the "norms and networks that enable people to act collectively" in a community (Woolcock & Narayan, 2000: 226). The role of social organisations, networks, and collaborative governance (or cooperation) prior to and post-disaster event can be key in a community's ability to respond to disaster events (Cretney, 2016). While social capital assists in making collaboration more effective, collaborative governance of disaster management is vital. Hence, the majority of this chapter focuses on the collaborative governance processes in drought management and response made possible through existing social capital networks.

R. Light (✉)
Department of Geography and Environmental Studies, Stellenbosch University, Stellenbosch, South Africa
e-mail: lightrebeccann@gmail.com

© The Author(s), under exclusive license to Springer Nature Switzerland AG 2023

R. Donaldson (ed.), *Socio-Spatial Small Town Dynamics in South Africa*, GeoJournal Library, https://doi.org/10.1007/978-3-031-37142-4_4

The Eastern Cape experienced a severe drought between the years 2015 and 2022, which impacted water security and local economies, particularly within the predominantly rural areas, under-resourced communities, and impoverished areas (Mahlalela et al., 2020). Many of the small towns in the Eastern Cape, including Makhanda (Grahamstown), Graaff-Reinet, Bedford and Queenstown, experienced significant water shortages during this period (Mahlalela et al., 2020). This chapter examines the role of collaborative governance through existing social capital in disaster management during and after this severe drought period by studying the case of the small town of Graaff-Reinet in the Eastern Cape, South Africa.

A collaborative governance regime (CGR) emerges where government institutions collaborate with non-state stakeholders (such as community members, forums, businesses, or NGOs) to solve a common problem. The collaboration in Graaff-Reinet was possible through the vessel of the new, community-established Graaff-Reinet Economic Development Forum (GEDF). Through this collaboration, vital research was done on water supply and infrastructure in the town, which led to the immediate injection of water into the supply network, as well as the possibility of a large-scale water infrastructure project. This project, the Nqweba Dam dredging project, could not only ensure water security for the town during future drought periods but also potentially lead to similar projects across the water-scarce country. Thus, this study thus brings the focus of social capital in disaster management away from immediate disaster responses (which currently dominates social capital disaster literature) and into the realm of long-term disaster planning. However, the failure or success of such a long-term collaboration between a community and government relies heavily on the strength and sustainability of the CGR. This case study analyses the Graaff-Reinet CGR by applying the CGR conceptual framework by Emerson and Nabatchi (2015) using qualitative data in the form of interviews with key stakeholders and secondary resource data (such as research reports conducted for stakeholders and a local news review), to determine the success and sustainability of this CGR in responding to drought and preparing for future drought events. First, it is important to understand the concepts of social capital and collaborative governance and how they were applied in the context of the case study.

4.2 Social Capital

Social capital can be understood as the resources (human or otherwise) that can be garnered from social organisations (networks, norms and social trust) by either the individual or the collective (Bartkus & Davis, 2009). The networks, institutional relationships and attitudes that make up "social capital" facilitate community coordination for mutual benefit, or at times a common goal regularly referred to in social capital literature as "the common good" (Putnam, 1995; Grootaert & van Bastelaer, 2002; Bartkus & Davis, 2009). Community coordination can sometimes come in the form of collaborative governance. One of the most frequent criticisms of social capital is the difficulty in defining the concept. That is, definitions vary across seminal authors, contexts, and applications. Though the chameleon-like nature of social capital can be seen as a weakness in terms of designing and replicating research approaches, the way that social capital changes according to context need only be understood so that the correct approach is applied in the correct context.

Though the three considered seminal authors of social capital, Pierre Bourdieu (1986), James Coleman (1988) and Robert Putnam (1995, 2000), have different approaches to the concept, there are three consistent principles that remain throughout. Firstly, social capital is created and exists within social interactions and connections between people (made stronger through repeated interactions and norms). Secondly, that value is gained from these connections, and lastly, that a level of trust must exist for this exchange of value to occur. What made Putnam distinct from Bourdieu and Coleman, was his focus on how social capital benefits the greater community (Putnam, 2000: 2). He defines social capital as referring "to features of social organisation such as networks, norms, and social trust that facilitate coordination and cooperation for mutual benefit" (Putnam, 2000: 2). This definition is far more condensed than the works of Bourdieu and Coleman and encompasses all the most important pillars of social capital: *networks*, *norms*, and *trust*. All these combine to create an environment that allows civil society to take advantage of the capital created to reach a common goal. In terms of social networks or structures, Putnam focused on networks of civic engagement in his study of subnational governments in different Italian regions. He found that the prevalence of such networks, if strong, fostered "sturdy norms of generalised reciprocity and encourage the emergence of social trust", which then ultimately "allow[ed] dilemmas of collective action to be resolved" (Putnam, 2000: 2–3).

While Bourdieu and Coleman focused on person-to-person direct interactions, Putnam took the concept further and widened the lens to observe whole communities *and* governing powers and the collective good that social capital within this context could help produce. This is where types of social capital need to be defined. It is popular to divide social capital into three main categories: bonding, bridging, and linking social capital.

Bonding social capital can be defined as "strong intense ties within group trust and norms of reciprocity" (Besser & Miller, 2015: 176). It describes the network connections that are usually found within a homogenous group (Bhandari & Yasunobu, 2009: 501; Claridge, 2018: 1). Within the disaster literature context; it is common that one of the outcomes of bonding social capital is that neighbours tend to assist each other during and following disaster events (Hawkins & Maurer, 2010: 1780). Though bonding has stronger ties compared to bridging, they typically have a weaker by-product (Hawkins & Maurer, 2010: 1780).

Bridging social capital refers to the 'loose ties' found across organisations and communities, closing the gap between different groups within a society, and it is more 'open membership' than bonding social capital (Emery & Flora, 2006: 21; Bhandari & Yasunobu, 2009: 501). As the name suggests, bridging social capital acts as the 'bridge' between organisations and communities (Rogers & Jarema, 2015: 20). Though the ties in these types of networks are 'weaker' in comparison to bonding social capital, the outcomes tend to be favoured over bonding social capital in literature, and the negative effects of bridging social capital are few if there are any (Coffé & Geys, 2007: 124; Claridge, 2018: 4). The weaker ties make participation in the networks, and thus the actions of the networks, voluntary: which means that the door is left open for change without negative social sanctions (Claridge, 2018: 3).

Linking social capital is an extension of bridging social capital. Szreter and Woolcock (2004: 665) define linking social capital as "norms of respect and networks of trusting relationships between people who are interacting across explicit, formal or institutionalised power or authority gradients in society." Linking social capital connects community groups and individuals to institutions with power or in important social positions, such as government institutions (Bhandari & Yasunobu, 2009: 501; Hawkins & Maurer, 2010: 1779). The nature of linking social capital allows civil society to connect with government and, under the right circumstances, this could improve government performance and service delivery. People and groups have "multiple and overlapping relationships" with each other, and one cannot examine a community as if only either bridging, bonding, or linking social capital exists (Claridge, 2018: 2). Activities where civil society and public entities interact with one another to solve a common problem go further than these parameters of social capital. Though social capital alone can assist with providing short-term solutions, managing drought impacts and preparing for future droughts requires some form of governing.

Traditionally, the task of drought preparedness and management would fall on governments and government entities. However, it is often the case in South Africa that governments lack the capacity or political will to engage in long-term anti-drought effect efforts (Makaya et al., 2020). In fact, according to Tierney (2012: 342) "the concept of governance itself arose in part from the recognition that functions that may formerly have been carried out by public entities are now frequently dispersed among diverse sets of actors that include not only governmental institutions but also private sector and civil society entities." Collective action in the face of an acknowledged community problem that requires long-term planning and solutions is more likely to be achieved through collaborative governance than the looser networks of social capital alone, though existing bridging and subsequent linking social capital networks can allow for collaborative governance to form.

Collaborative governance describes a process of decision-making and management that involves public agencies and non-state stakeholders. Ansell and Gash (2008: 544) define it as "a governing arrangement where one or more public agencies directly engage non-state stakeholders in a collective decision-making process that is formal, consensus-oriented, and deliberative, and that aims to make or implement public policy or manage public programs or assets." One of the key focuses of this definition is that such a forum is instigated by public agencies. Emerson et al. (2012: 2) define it as "the process and structures of public policy decision making and management that engage people constructively across the boundaries of public agencies, levels of government, and/or the public, private and civic spheres in order to carry out a public purpose that could not otherwise be accomplished." The Ansell and Gash definition emphasises that collaborative governance is instigated by public agencies, while Emerson et al. emphasise that it involves actors from various spheres, including public, private, and civic. Social capital can affect collaborative governance at different levels. Existing relationships or social capital networks are crucial as a communication network to suggest a collaborative governance effort already exists. Actors in a bridging social capital network will already know what other actors can bring to the table in terms of knowledge and resources (Oh & Bush, 2016). Additionally, loose networks of social capital, like bridging or linking, can improve this process by expanding the flow of resources and knowledge (Oh & Bush, 2016: 229).

As will be found later in this chapter, the collaborative governance processes that are being observed were initiated by private and civil parties – though participation by public agents allowed for it to function. The Emerson et al. (2012) definition also mentions that public, private and civic spheres operate together to achieve an identified public purpose as opposed to collaborative governance existing only to manage a single program or asset. This links back to social capital theory and is more aligned with how such governance would work in the context of drought management. Collaborative governance can also refer to a framework (Voets et al., 2021: 1). The conceptual framework for a collaborative governance regime (CGR) by Emerson et al. (2012) shows the different drivers, actors and factors involved in the collaborative governance process (Fig. 4.1).

This conceptual framework was used to study the Graaff-Reinet Economic Development Forum (GEDF), the entity identified that initiated a collaborative

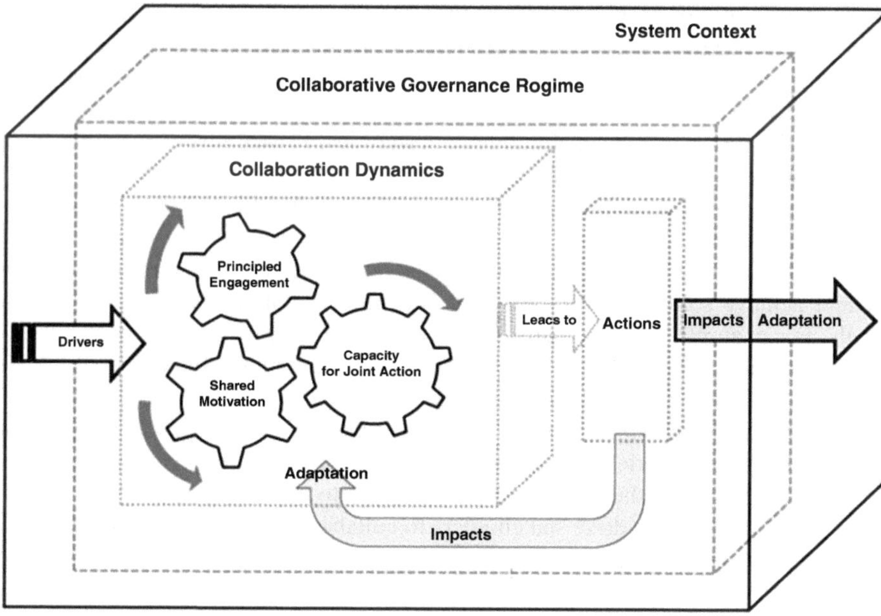

Fig. 4.1 Conceptual framework for a collaborative governance regime. (Emerson et al., 2012: 6)

governance regime and the public entities from the Dr. Beyers Naude Local Municipality (DBNLM) involved in drought planning and water security planning in the case of Graaff-Reinet.

4.3 Collaborative Governance Regime Framework

Figure 4.1 is a visual representation of how a collaborative governance regime would operate. Firstly, the regime operates within a unique system context. This encompasses all the outside influences that either affect or are affected by the CGR. This can include the political environment, socioeconomic factors, the environment itself, policies, authorities, power relations, historical context, previous conflicts or collaborations, diversity, cultural climate, trust levels, and degree of connectedness within and across existing networks (Emerson et al., 2012).

There is a difference between the initial conditions present when collaborative governance emerges that facilitate or discourage cooperation and the essential *drivers* of a CGR (Emerson et al., 2012: 9). The latter are factors that lead a CGR to be officially formulated at the beginning of the CGR process. Drivers include factors of leadership, consequential incentives, interdependence, and uncertainty (Emerson et al., 2012). Leadership, the first essential driver of a CGR, would refer to a person or organisation that initiates the CGR through securing resources (financial, human, social) and garnering support for the initiative. For stakeholders to be pushed to act together, there are usually *consequential incentives*. Negative incentives usually come in the form of problems that are time-sensitive or issues that, if not handled, will lead to negative impacts on stakeholders or the community. A disaster event like a drought, for example, leads to damage to livelihoods or water insecurity in a community- which would incentivise stakeholders to act. Positive incentives can come in the form of grants or funding opportunities. Private investors with a stake in a community that are affected by a particular problem could incentivise a CGR by promising funding for their undertakings- an example of both positive and negative incentives.

Interdependence occurs when individuals, organisations or government entities cannot achieve something on their own and need to collaborate to achieve their goals. For example, a municipality may lack the funding to manage public infrastructure (dam walls, pipes, wells) and private entities with the funding do not have authority to manage the infrastructure themselves. Due to the nature of water policies in South Africa, situations regularly arise where the interdependence between private and public entities are obvious in terms of providing water to communities (Visser, 2015). When there is uncertainty of how to address a problem or how to define a problem, it can drive stakeholders to collaborate in doing so. Where uncertainty cannot be resolved, stakeholders can collaborate to "reduce, diffuse, and share risk" (Emerson et al., 2012: 10).

The innermost box of the CGR framework is labelled as *collaboration dynamics* which consists of three interactive components: principled engagement, shared motivation, and capacity for joint action. These three components interact to produce actions. These actions can produce impacts within the regime as well as outside of the regime, hence the arrows linked back towards these three principles in the diagram. Through the process of principled engagement, the different stakeholders coming from different positions communicate across their institutional, sectoral, or jurisdictional boundaries to solve the identified problems, resolve conflicts or create value (Emerson et al., 2012: 10). The engagement is principled as it involves "civil discourse, open and inclusive communications, balanced representation of all relevant and significant different interests" and is "informed by the perspective and knowledge of all participants" (Emerson et al., 2012: 11).

Shared motivation links strongly to social capital in that it "highlights the interpersonal and relational elements of the collaborative dynamics" (Ma et al., 2018:358). It consists of four elements: mutual trust, internal legitimacy, and commitment. Trust generates mutual understanding, which generates legitimacy, which leads to bonds of shared commitment. The idea of *trust* and how it functions in a network has been discussed in the literature on social capital. *Mutual understanding* occurs when stakeholders understand and respect one another's positions and interests (whether they agree or not). *Legitimacy* occurs when participants in the CGR are found to be trustworthy, credible, and compatible in terms of interests. This both motivates and legitimises the collaboration to continue. Reciprocity and subsequent trust through this process can create bonds of commitment. *Commitment* allows participants previously separated by different boundaries to act collectively towards a shared goal.

Collaborative dynamics must function to generate the capacity for joint action that did not exist prior to the establishment of the CGR. To determine the level of capacity for joint action that has been generated in a CGR, the four elements of the capacity for joint action were assessed: leadership, knowledge, institutional arrangements, and resources.

Leadership was already established as a driver of CGRs, but it can be an outcome of collaborative governance as well in that new opportunities for leadership roles are cultivated. In practice, that could be observed through the election of a representative for a newly created forum, a convenor for engagements or a public advocate. *Knowledge* in this context refers to "the social capital of shared knowledge that has been weighed, processed, and integrated with the values and judgements of all participants" (Emerson et al., 2012: 14). For example, experts that were part of the process could share the data they gathered during the process, and it could be integrated into future plans. *Resources* refer to funding, time, technical support, administrative assistance, analysis skills, expertise, or even power. All of these and others would be vital in creating the capacity of joint action. Power is more nuanced and cannot be recorded by the researcher as easily as the other concrete resources. Therefore, interviews and observation were necessary to determine how power is distributed in the CGR.

Institutional arrangements would include all the process protocols and organisational structures implemented to repeat interactions between stakeholders over time. For a CGR to be sustainable, more permanent/explicit structures and protocols need to be established. Arrangements can be both intraorganizational (how a group manages itself) and interorganisational (how all participants manage together and integrate decision-making). Different stakeholders or groups are given different functions and can take the form of "a self-managing system", a "designated lead agency", or a newly created governmental structure (Emerson et al., 2012, 14). This falls under the social capital network identification part of the research process, as well as the observations that will determine the strength of the CGR itself.

The main purpose of a CGR is to provide mechanisms for the collective action originally determined by stakeholders and actors in previous stages of the process within the CGR. These actions can be conducted by the CGR members (individually or collectively) or by external entities in response to direction or recommendations by the CGR (Emerson et al., 2012: 17). The actions are distinct from the impacts. For example, an action could be the implementation of a new policy, and the impact would be how the policy changes things on the ground. Impacts can be intentional and unintentional, relevant to the original goals or not. They can also be both negative and positive. If targeted outcomes are specified during the collaborative dynamics processes and derived from a shared theory of action, impacts are more likely to be in line with them (Emerson et al., 2012: 18).

4.4 Case Study: Collaborative Governance Regime in Graaff-Reinet

The case study applied the CGR framework discussed above to both determine the outcomes of the collaboration between the GEDF and the DBNLM, as well as the sustainability of such a relationship within the context of drought planning and long-term water security. As such, the system context and drivers of the CGR supplied background information on the study area, analysis of the collaboration dynamics look at the establishment and functioning of the CGR, and the actions, impacts, and adaptations of the CGR will determine the outcomes of the collaboration. The overall sustainability of the CGR and that value of the case study are covered in the discussion of these factors.

4.5 System Context

Graaff-Reinet is a small town found in the Great Karoo area in the Eastern Cape, South Africa. As one of the oldest towns in the country, it is considered a significant town in the place of South Africa's history. Besides farming, tourism forms a large part of the economic activity in the town (Fig. 4.2).

The most recent census data shows that Graaff-Reinet had a population of 35,672 people in 2011 (Frith n.d). That includes the greater Graaff-Reinet area, which is

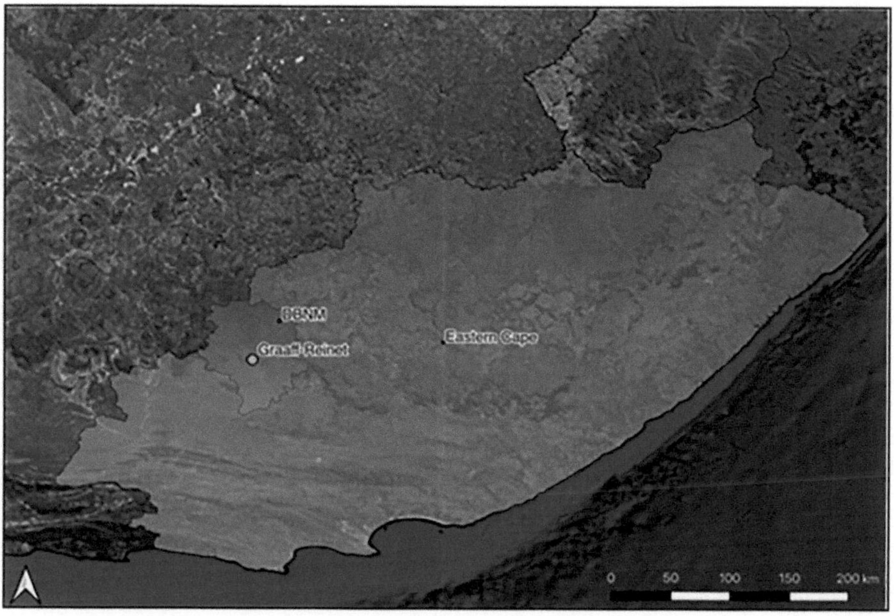

Fig. 4.2 Position of Graaff-Reinet within the Eastern Cape and Dr Beyers Naude Local Municipality (DBNLM). (Source: Author 2022)

Table 4.1 Demographics of Graaff-Reinet (Frith n.d.)

Suburb	% Coloured	% White	% Black	% Other
Adendorp	16.96	78.05	4.24	0.75
Asherville	91.01	0.06	8.40	0.54
Graaff-Reinet SP	20.75	52.29	25.58	1.37
Kroonvale	95.01	0.09	3.92	0.98
Spandauville	34.67	29.36	34.94	1.03
uMasizakhe	19.28	0.03	80	0.68

made up of Adendorp, Asherville, Graaff-Reinet SP (or the 'Horseshoe'), the Karoo Nature Reserve, Kroonvale, uMasizakhe and Spandauville. Table 4.1 shows the demographics for each suburb and Fig. 4.3 shows each suburb in relation to the data.

In terms of the political climate, the town has seen significant changes prior to and during the drought period. In 2016, the Camdeboo Municipality responsible for Graaff-Reinet was incorporated into a newly amalgamated Dr. Beyers Naude Local Municipality. At this stage, this new municipality had yet to implement a disaster management plan, leaving the government unprepared for a disaster event such as the drought. A larger area of governance also meant that the municipality's capacity became stretched, and debts of smaller municipalities that had been incorporated into the DBNLM put financial pressure on the new municipality.

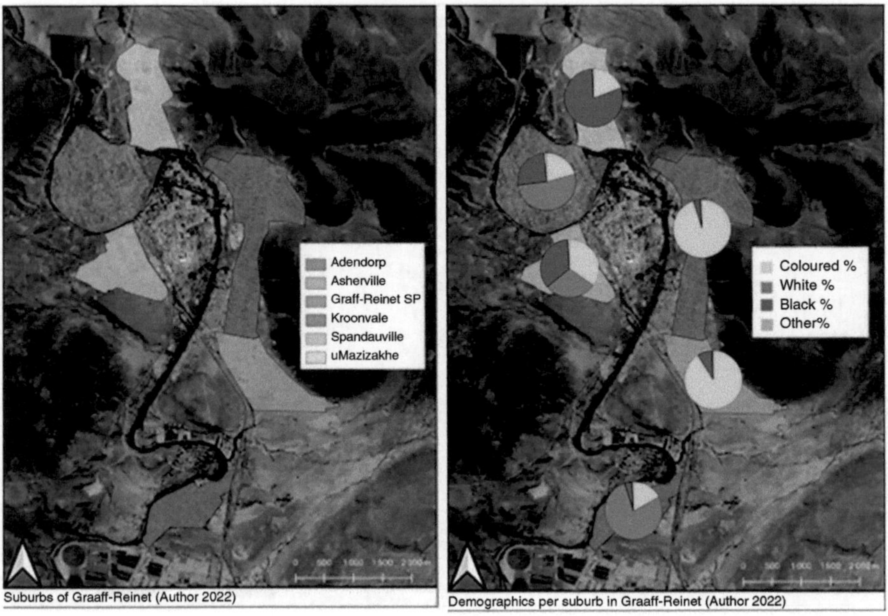

Fig. 4.3 Layout of suburbs in Graaff-Reinet compared with demographics. (Source: Author 2022)

Graaff-Reinet is a water resource-dependent town in a semi-desert climate. The foremost water source for Graaff-Reinet prior to the drought was the Nqweba Dam (shown in Fig. 4.4), which was already experiencing reduced capacity duc to silt build-up. After construction, the dam had a capacity of 78.8 million m^3, but due to silt build-up, it is currently only at 42% of its capacity at 33 million m^3 (Basson et al., 2022). When the dam runs dry, the town is otherwise dependent on borehole water, which itself is depleted when there has been little to no rainfall (Spocter, 2011).

4.6 Drivers

As was discussed above, drivers of a CGR include factors of leadership, consequential incentives, interdependence, and uncertainty. Consequential incentives that drove the initial formation of the CGR were related to the prevailing drought and the impact thereof. To provide further background in the context of the drought that began in 2015, Fig. 4.5 below is a summary of the timeline of events leading up to the involvement of the GEDF compiled using local newspaper headlines of the Graaff-Reinet Advertiser.

Observing the above, we see that though concern of water security was present as early as 2016, the drought and water-related issues only began to dominate local

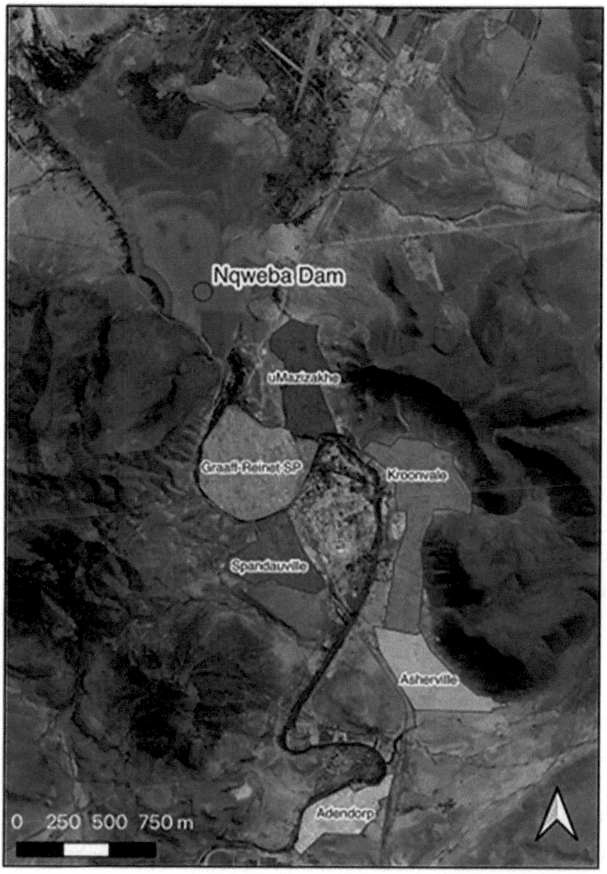

Fig. 4.4 Nqweba dam in relation to Graaff-Reinet. (Source (Author, 2022))

news headlines in 2019 at the height of the drought when the Nqweba dam had officially run dry (see Fig. 4.6).

The communication between the DBNLM and the community had almost completely broken down by 2019. This was made clear in two instances: water truck deliveries, and safety of tap water. In terms of the former, community members were complaining that they were not made aware of delivery schedules, meaning that sometimes they missed getting their share of the delivery. This also led to fights over water when trucks did arrive and took residents by surprise. In terms of water-safety concerns, a rumour stating that water from the taps was no longer drinkable gained traction to the point where the local government had to officially address the issue. The concern was that the water from the Nqweba Dam was toxic due to low levels. The municipality responded by saying that they were no longer even making use of the Nqweba Dam to provide the town with water and had switched to boreholes. The fact that this had not been communicated in the first place showed an

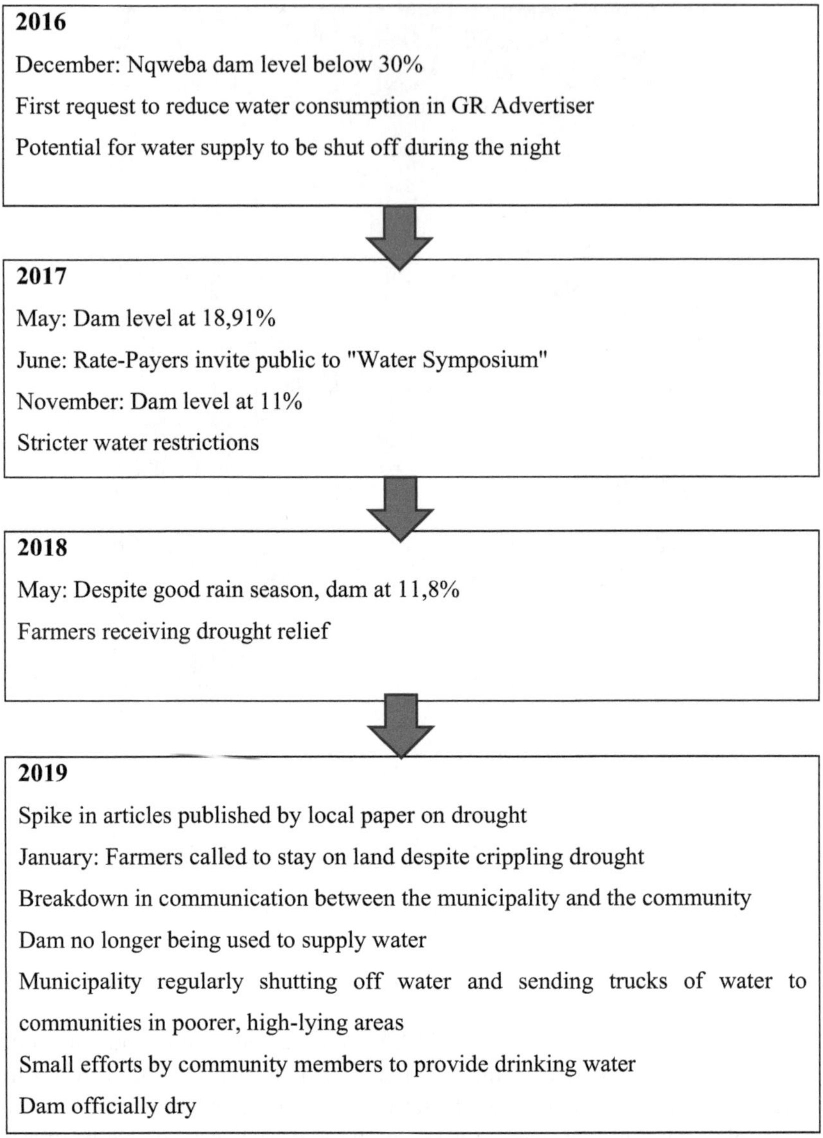

Fig. 4.5 Summarised timeline of events leading to the establishment of the GEDF and collaborative governance regime in Graaff-Reinet

extreme disconnect between the local municipality and the community about the dire circumstances of the drought.

Despite immediate disaster responses in the form of water trucks and NGOs making efforts to donate drinking water to those who could not afford to buy water, it soon became clear that more was needed to respond to the water crisis the town was facing. The local municipality struggled to provide the town with water due to

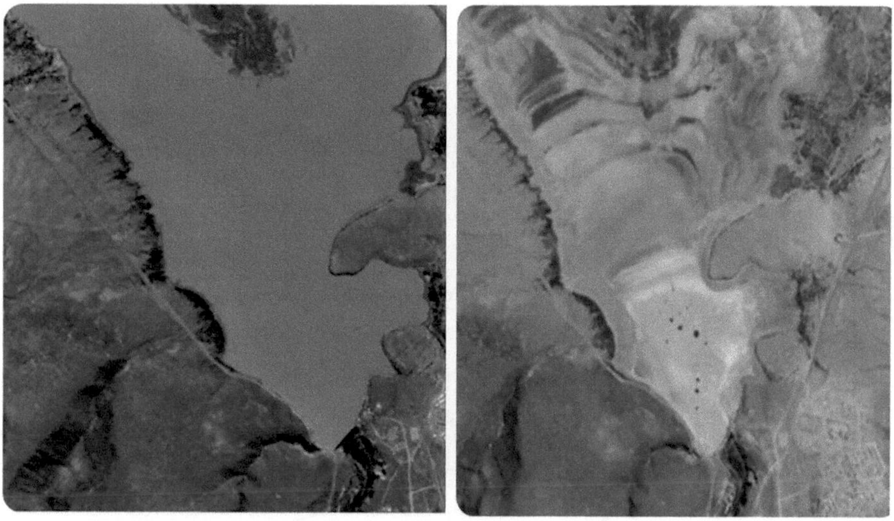

Fig. 4.6 Nqweba Dam when full (June 2009) versus Nqweba Dam in September 2019 (Google Earth)

a breakdown of water-related infrastructure not connected to the dam due to poor maintenance and lack of use over recent years. A disaster management plan for the town had not been formulated since the amalgamation of the municipality in 2016, and it was becoming apparent that the municipality lacked the capacity to respond to the drought event. Their lack of capacity extended beyond drought response and garnered the attention of local community members. This is an example of interdependence- the local government needed assistance, and community members lacked the decision-making power to make a significant impact.

In 2019, a local community leader and attorney (Interviewee A) was approached by concerned community members about the economic state of the town and various issues that needed attending to, which the municipality had neglected due to lack of capacity. These issues included poorly managed dumpsites, poor service delivery, and sewerage infrastructure damages. Though some community members suggested legal action, Interviewee A wanted to approach the community with an alternative plan. At a public meeting hosted by Interviewee A, community members from various interest groups decided to form an approach to solving the community's problems by partnering and communicating with the local government. Following this meeting with local businesses, community members, and interest groups (the Rate-Payers Association, the Chamber of Commerce, the Black Businessman's Forum) the Graaff-Reinet Economic Development forum was established. The forum's goal was to collaborate with the local government to enhance its capacity to improve Graaff-Reinet's circumstances. This is an example of how leadership drove the initiation of the CGR. Through existing connections between community members and interest groups and the leadership of Interviewee A and others, the forum allowed increased communication between the municipality and the

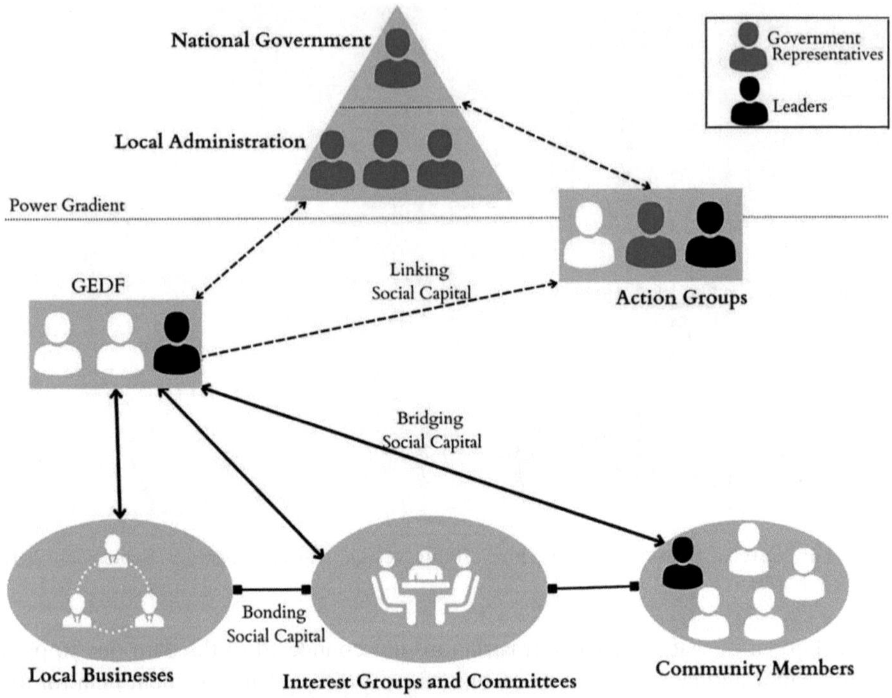

Fig. 4.7 Summarised diagrammic model of social capital in Graaff-Reinet. (Author, 2022)

community. The diagram in Fig. 4.7 above is a visual representation of the social capital networks involved in this context.

The above shows that, in terms of social capital, *bonding social capital* between community members led to *bridging social capital* with the GEDF, which allowed for *linking social capital* with the municipality. Trust, a vital component for social capital networks to exist, had been broken down between the community and the municipality. The GEDF, formed between community members with higher levels of trust, meant that trusted community leaders could establish a link with the local municipality. The power gradient is relevant here, as government entities have more decision-making powers within this context of water provision, water security and drought preparedness. The community, comprised of private businesses with physical capital and individuals with human capital in the form of expertise and skills, was theoretically able to assist the local government experiencing decreased capacity. Therefore, both the GEDF and DBNLM needed to collaborate to solve the water and drought-related problems of the town. Once communicated by the GEDF, this shared goal allowed collaboration to form between the community and local government. From the first collaborative undertaking between the DBNLM and the GEDF, the beginnings of a collaborative governance regime began to materialise.

The impact of the drought on the community influenced the GEDF to make water security one of their priorities, with the focus being on long-term solutions to water insecurity and ensuring that likely future drought event impacts are mitigated against. An outreach session was hosted by the DBNLM on the 11th of October, where the GEDF communicated their desire to collaborate with local government on addressing drought and water security issues. Uncertainty on how to approach the complex problem of water security in the town drove the formation of a focused, collaborative action group. Following a meeting with local municipality representatives in November of 2019, a Joint Operations Committee (JOC) was established to this end- where members of the GEDF collaborated with members of the municipality to address water security issues. The JOC had representation at local, central, district, and provincial government level, as well as civic organisations. The GEDF committed to assisting the DBNLM in conducting research on water supply issues in Graaff-Reinet. It was from this point on, that the study sought to observe the collaborative governance regime that had emerged in terms of its functioning, impacts, and sustainability.

Though the GEDF itself is not a collaborative governance regime, the connection and communication across network channels between it, local government, and eventually national government presents as such. The impacts and actions of the GEDF and the JOC that were identified would determine the success of the collaboration. However, sustainability of the relationship is equally relevant when attempting to predict success in the long-term, which is why it is necessary to assess factors such as principled engagement, shared motivation, and capacity for joint action.

4.7 Collaboration Dynamics: Engagement, Motivation, and Capacity for Joint Action

As was stated above, collaboration dynamics in a CGR can be examined through principled engagement, shared motivation, and the capacity for joint action. These factors were observed between 2019 and 2022 through attending meetings, tracking activities through secondary resources and publications, and finally through interviews with key stakeholders in 2022. A linear timeline approach was deemed necessary as external factors such as the COVID 19 lockdown had a significant effect on collaboration dynamics between stakeholders, and resistance against such influences added another dimension when observing the sustainability of the CGR.

4.7.1 Emergence of a CGR: Initial Assessment

Initial trust was formed between the GEDF and local government following the commitment of the GEDF to assist in providing the municipality with specialised research on water supply and water security-related solutions, which was timeously and expertly delivered. The municipality reciprocated by encouraging the formation of the technical sub-committee of the JOC which was to make use of these reports and address issues of water supply in the town. The GEDF served on this technical sub-committee and assisted in arranging a Water Summit in December 2019. This summit provided a workshop-like space to collaboratively identify the problems facing the town in terms of water security, as well as to align objectives and interests across institutional levels. For a CGR to be considered effective, aligned objectives and compatibility among stakeholders is vital. The summit assisted in this, though repeated meetings were necessary to ensure continued alignment and commitment across levels. This is where principled engagement became vital.

Principled engagement involves communication, representation and informed perspectives. Initially, communication between the GEDF, JOC, and DBNLM was strong. Enthusiasm following the Water Summit drove regular meetings and communication. Engagement was also legitimised by a widely representative collaboration. In terms of the GEDF, the interests of community members were broadly represented through local businesses, the Black Businessman's Forum, the tourism organisation, Chamber of Commerce, the District Agricultural Union, Rate-payers Association, the Community Works Programme, the Siyanzela Education Programme, and several local leaders of different races and cultures. The local municipality's needs were accounted for through their representation on the JOC. In terms of informed perspectives, this was achieved through initial meetings and scientific reports provided to stakeholders (to be discussed later in the chapter).

Though the CGR initially seemed well-equipped in terms of shared motivation and engagement, capacity of joint action needed to be present. Leadership, knowledge, resources, and institutional relationships are the factors that were observed to determine this initial capacity. There was clearly defined leadership in the GEDF where Mr. Interviewee A, who was instrumental in the establishment of the forum, was elected chairperson. The JOC was led by Interviewee B, who represented the local municipality. Mr. Interviewee A drove the collaboration in his capacity as chairperson of the GEDF. Institutional relationships within the GEDF were clearly demarcated in their constitution. However, the CGR itself had no clear institutional arrangements beyond the JOC. Structures appeared only as permanent as their members, which does not bode well for sustainability. At this stage the CGR was in its relative infancy and there was little funding for permanent employees or a succession training plan.

The values of the GEDF were clearly demarcated in their constitution. For example, one of the objectives of the GEDF is stated in their constitution as:

> To promote interaction between local government, the community, civic organisations, public benefit organisations, business and civil society with a view to establishing partnerships and joint adventures to provide for the basic needs of all members of the community through effective service delivery. (Constitution Graaff-Reinet Economic Development Forum, 2020: 4).

The stakeholders in the CGR representing the municipality aligned with this value as they communicated their willingness to be part of such interactions. However, in terms of politics, the GEDF clearly presented itself as a-political, while the municipality was often influenced by political motivations (Interviewee A, 2021). Though this was not a barrier in the context of water security issues, other side projects by the GEDF motivated for the betterment of the community were challenged on occasion by municipal representatives for political reasons. For example, the GEDF assisted a feeding scheme in a community where the ANC counsellors had influence without input from the municipality. Those in the DBNLM aligned with ANC became angry at the GEDF's interference in what they deemed their territory, despite not initiating or intending to initiate a feeding scheme themselves (Interviewee A, 2021). It became clear that the stakeholders representing the GEDF had to be cautious when it came to service provision or projects in the town, and they adjusted by ensuring that "whatever good came from the collaboration, the credit went squarely to the municipality" (Interviewee A, 2021).

In terms of shared knowledge, the GEDF secured funding from the private sector for scientific research on water supply in Graaff-Reinet, further showing their commitment to solving the water security crisis faced by the town. This research become the most valuable output of the collaboration. Table 4.2 below summarises the reports provided to the municipality and GEDF:

Table 4.2 shows that extensive specialist knowledge was provided and presented to all relevant stakeholders through the efforts of the GEDF. The first report in 2019 and the two September 2020 reports assisted the DBNLM in decision-making and planning when it came to water provision and infrastructure in the town. It allowed them to identify problems that needed immediate address, such as damaged infrastructure. Additionally, the reports assisted in identifying sites for boreholes and wells to address immediate water supply issues. One of the more significant outcomes of the multitude of reports funded via the GEDF was the identification of long-term solutions to water security: vital for a town likely to experience increased frequency and severity of drought in the future. Of the proposed solutions, a Nqweba Dam Dredging project was identified by Basson et al. (2022) as the most feasible solution to decreased dam capacity and, consequently, long-term water security in Graaff-Reinet.

This proposed project became the focus of the GEDF, while the municipality still attended to immediate water crisis mitigation. The Nqweba Dam dredging project proposal by Dredging Africa gained the attention of foreign interested parties (Interviewee, 2021). Through the Dutch company Royal IHC, the government of the Netherlands reached out to the GEDF and communicated that they would provide a portion of the funding for such a project if the rest of the funding could be secured by a fixed date in 2021. The GEDF presented this to the municipality and the Department of Water and Sanitation, who indicated that though they supported the project in theory, they would not be providing funding.

While the GEDF was attempting to garner support and interest for the dredging project, the municipality was facing a new challenge: the COVID-19 pandemic. The DBNLM redirected time and resources towards dealing with the fallout of the pandemic and the lockdown period. When representatives of the municipality did meet with the GEDF, discussions and initiatives centred on dealing with this fallout. Though the GEDF offered assistance in this period (not the focus of this study), the

Table 4.2 Summary of water-related reports organised by GEDF through private funding[a]

Report title	Date	Summary
Graaff-Reinet: Water resources, supply, distribution and water infrastructure: Abbreviated proposal: investigation, medium- and longer-term planning	November 2019	Water demand projections, optimal use of water resources and further development, groundwater recharge options, development of water masterplan, options for re-use of wastewater, possible effects of climate change and interventions, risk analysis, and firm yield of Nqweba Dam.
Nqweba Dam, Graaff-Reinet: Project Plan and Proposed Cost Structure –	March 2020	Report regarding possibility of a sedimentation dredging project for the Nqweba Dam.
Memorandum on background information for dredging of the Nqweba Dam	April 2020	Background information for possible dredging operations of the settled sediment of the Nqweba Dam and description of the status quo of sedimentation in the Nqweba Dam reservoir, historic water level statistics, and a description of a provisional dredging configuration.
Feasibility study Investigation of Sedimentation Mitigation Measures: Nqweba Dam at Graaff-Reinet	July 2020	Reviews the historical sedimentation of the Nqweba Dam and also to investigate suitable mitigation measures to deal with sediment to increase the firm water yield of the dam. Presents: status quo of sedimentation in the reservoir, historic reservoir water level statistics, water requirements for GR and estimated up to 2050, current water resources of GR (ground water and dam), hydrology of the dam, dam catchment sediment yield, historical firm water yield calculation, and possible reservoir sedimentation mitigation measures.
Hydraulic model analysis of the existing Graaff-Reinet water network	September 2020	Compilation of a hydraulic model and analyse the existing water network for GR located in the DBNM.
Study to analyse treasury data and identify projects that promote WC/WDM in GR	September 2020	Analysis of the treasury data in order to identify billing and water meter reading anomalies, and to identify projects that could be implemented to promote WC/WDM in order to reduce non-revenue water, which includes water losses.

[a]Resources provided to author by the Graaff-Reinet Economic Development Forum

leadership of the GEDF was still concerned with the drought. As compatibility and alignment between the GEDF and the municipality began to slip, vital regular communication was severed due to the prevailing pandemic, which paused in-person meetings. During this period, the initial enthusiastic private funding that allowed the GEDF to fund the above research began to diminish as local municipality damaged trust with private investors through unrelated political endeavours and detrimental media coverage of municipal financial movements (Interviewee A, 2021; Interviewee B, 2022). The deadline to provide a portion of the funding of the project came and went, and during 2020 the CGR seemed to have all but dissipated. What started off as a strong collaborative relationship evaporated in the face of external challenges and the withdrawal of much of the initial private funder's support. Without reciprocated trust, one of the most vital drivers of a successful CGR, the collaboration diminished.

4.7.2 Revival: Testing the Resilience of the CGR

In 2022, the influence of COVID-19 restrictions on the regular convening of the GEDF lessened. Leaders within the GEDF and specialists who had spent time on the Dredging Project planning decided that the potential of such a venture was too important to abandon.

> We reinitiated discussions with all the role-players, including international funding, but also arranged that our specialist team come up to re-present the project to the local and provincial government, as well as the Department of Water and Sanitation. We sought to present the project properly so that all fears of the different role-players could be dispelled. (Interviewee, 2022)

After the project was re-presented, these role-players provided their full support once again. With new vigour and re-opened channels of communication, the GEDF revisited the funding process with the goal of being less dependent on private investors. The new commitment of local role-players reignited interest from international funders who presented new deadlines for partial funding for later in the year 2022. During this period, the new Minister of Water and Sanitation, Senzo Mchunu, bought in to the idea of the potential of de-silting projects after a report was released in January on de-silting in South Africa where the Nqweba Dam was identified as one of the possible pilot zones for dredging dams in South Africa (Water Resource Management, 2022).

> This cast a new light on what we were doing. We have on-going engagement now with DWS, environmental affairs and local authority. The challenge now is to get 25% of funding from private sector, as well as funding for administration requirements needed to make funding available for government. (Interviewee A, 2022)

The support from local government was re-established, and collaboration between local, provincial, and even national government began anew. Stakeholders representing the Dutch government and the interested Dutch company contacted the GEDF and agreed to meet in South Africa with them, public representatives, and specialists to re-assess the feasibility of the project. This meeting was met with enthusiasm and support from all stakeholders, and trust between the GEDF, foreign interest parties, and the government was built up again. With the passing of the COVID-19 lockdown and the worst of the drought, the focus of the collaboration became the potential dredging project. This clear objective allowed the collaboration to be stronger than it had been initially.

However, the political climate of Graaff-Reinet presented challenges. Following the lack of a clear majority in newly elected municipal council, the DBNLM was saddled with an ineffective council for months. An added impact of this was that an IDP for 2022/2023 has yet to be published- meaning the objectives of the DBNLM could not be officially stated or acted upon. Beyond the fact that this led to red tape in acquiring funding for the municipality, the discord discouraged private funding for projects involving the municipality. Despite new vigour amongst leaders in the GEDF and certain interest parties, the collaboration struggled to obtain the promising stability and legitimacy it had at its inception.

> COVID broke down our ability to engage for some time, which meant the executive [of the GEDF] would run with things. Stuff changes in your collaborative organisations. Political parties change and you work with constantly changing administrators. So, we are mindful now that we need to re-engage. This thing [collaboration] could die with us. (Interviewee A, 2022)

Here Interviewee A, who remained chairperson of the GEDF in 2022, expresses an important concern of sustainability. Firstly, he alluded to the fact that the executive of the GEDF would execute actions without input of either the GEDF or municipal collaborators during the breakdown period. While on the face of things, this might 'get things done,' it undermines the collaborative process and reduces chances for reciprocity and consequential trust and engagement. In an interview with Interviewee C, a member of the GEDF representing the Graaff-Reinet Chamber of Commerce, Interviewee C expressed his concern with this.

> The GEDF was a good idea, and the collaboration was a good idea. But, you can't just have one or two people communicating in a letter saying 'we have done this, and we will not do this' and think that that is representative of a collaborative effort. It also does not seem sustainable if there is no permanence or plan for succession. (Interviewee C, 2022)

Interviewee C raises concern of leadership playing too strong a role, diminishing collaboration. He goes on to address what Interviewee A touched on above: the issue of succession plan and permanence. Beyond the obvious, that a CGR needs to be collaborative and representative, it needs to be sustainable. With only one or two leaders solely *driving* the collaboration, despite buy-in of stakeholders, the danger of the project "dying with them" is a concern. Despite renewed excitement and buy-in by local municipality and other stakeholders in the collaboration, roles were no longer clearly defined, and the executive of the GEDF alone was largely responsible for continued collaboration. According to Interviewee C, permanent roles which allow for succession were necessary for such a CGR to persist.

However, Interviewee C acknowledged that this would only be possible through employing permanent members in forums like the GEDF, requiring funding. This alludes to another challenge facing the CGR during this period: funding. Resources are imperative to a functioning CGR. While skills, power, experience, expertise, and knowledge were not in short supply within the CGR, funding was lacking.

Despite new vigour amongst key stakeholders for the continuation of the collaboration between the GEDF and the local municipality to address water security issues in Graaff-Reinet, it was observed that the CGR faced significant challenges in achieving their initial goal of water security. However, it must be noted that poor performance within certain collaboration dynamics only suggests that it is less likely that the actions and impacts identified at the start of the collaboration will be achieved. This does not mean that positive impacts resulting from actions during the collaboration process cannot occur. Intentional or otherwise, it is important as a final step of determining the success of the CGR to identify impacts caused by activities performed within the collaboration period.

4.8 Actions, Impacts and Adaptations

The analysis above alludes to a number of actions undertaken through collaboration that have led to impacts both intentional and unintentional within the Graaff-Reinet area and beyond. The formation of the JOC, the 2019 Water Summit, the research funded by through the GEDF, and the presentation of a proposal for the Nqweba Dam Dredging Project were all actions completed through the CGR process. This section discusses these actions and their impact on the Graaff-Reinet community's drought-related problems.

The formation of the JOC led to increased communication between stakeholders, allowed for focused effort and resources to be directed towards water security, and led to the seminal 2019 district Water Summit. The transfer of skills, power, and social capital via relationships formed in the JOC allowed the municipality to circumvent red tape and access SANParks land to drill additional boreholes and thus respond to immediate water shortages (Interviewee B, 2022). A review of the borehole infrastructure for Graaff-Reinet identified problem areas, which were communicated to the government via the JOC. These problems, such as damaged infrastructure and pump stations, were promptly addressed in December 2019. This led to an immediate increase of water into the supply network- another example of a short-term positive impact during the drought period. Figures 4.9 and 4.10 below shows examples of the issues reported in the borehole audit, and Fig. 4.8 shows the meeting where the results of the study was presented to the municipality.

Increased communication also allowed the municipality to respond to community concerns surrounding water truck deliveries (Interviewee B, 2022). The Water Summit had a number of impacts. It led to an effort by government to educate communities on water-wise behaviour in the form of education campaigns, which were budgeted for and included in the DBNLM IDP in 2021 (DBNLM, 2021). Knowledge and feedback shared during the summit, as well as research funded through the GEDF on water supply and infrastructure in the Graaff-Reinet area, informed a Water Management Plan drafted by the DBNLM in partnership with specialists introduced to them through GEDF connections. Therefore, policy and planning that had not been in place previously was now available to local government. Attention was brought to the need for medium- and long-term solutions to water security during the summit, which drove research funded through the GEDF on the topic.

During an excavation of the Nqweba Dam sediment for this research, it was discovered that there was more water stored in the silt of the dam than expected (Smith, 2020). Additionally, through this research, the Nqweba Dam Dredging Project was flagged as a potential solution to water security in the town for up to 50 years (Basson et al., 2022). An unintentional impact of publishing and presenting this research was the interest of the new Minister of Water and Sanitation, Mr. Senzo Mchunu, in the potential for dredging projects across South Africa in answering

Fig. 4.8 GEOSS representative Neville Paxton presenting the findings to the Dr. Beyers Naude Local Municipality. (Source: Graaff-Reinet Advertiser, 2019)

water security problems (Interviewee A, 2022). On this, Interviewee A said the following:

> Even if we fail to tie down private investors [for the project] in the end, we feel that we have achieved a lot. Through this collaboration, research which will inform government on long term solutions to drought and water management has been achieved and because of how it developed, it has become relevant as a national solution to increase dam capacity. (Interviewee A, 2022).

Increased capacity of local government, shared knowledge, and a shift in policies and planning by local government make up most of the notable impacts through the collaboration and functioning of the CGR. Smaller achievements were made possible through information on water infrastructure provided to the municipality by the research funded through the GEDF, such as addressing infrastructure weaknesses, additional wells, and informed approaches to delivering water to the community. The long-term impact of the CGR can only be determined over a longer observation time. Regardless, an informed municipality and community makes for a more resilient community in the face of future drought events.

4.9 Conclusion

The collaboration between the community forum of the GEDF and the DBNLM led to increased communication across boundaries, shared knowledge, shared resources, and laid the foundation for continued collaboration. The CGR was made possible

Table 34: Mimosadale Wellfield – Summary of Boreholes with concerns and recommendations

Mimosadale Wellfield		
Red blocks = Urgent intervention required		
Yellow blocks = Important matters		
White blocks = Further investigation required		
BH Name	**Concern**	**Recommendation**
3224AD00399	Pump was lifted by 9 m, reported to have black slimy clay and tree roots. This has offset telemetry and cut-off switches; borehole is currently manually switched on and reaches switch off level within 45 minutes. When operating optimally, borehole can add 103.68 m^3/day. 50 % of this (a conservative approach for the severe drought) means an additional 4% to the total wellfield supply.	STS - the service provider that does the telemetry system needs to come to site and reprogram telemetry to automate borehole and put back online. After a presentation of these findings to the local municipality, the municipality instructed the service provider to bring borehole back into production in this week (10-13 Dec 2019)
EC-N13-013	Drill chips and clay present on dipmeter. Also pumping less (2.06 L/s) than 2018 recommended rate of 3.9 L/s (for 24-hour pump cycles) – possibly as a result of sediment causing blockages. Measured water level was however below critical water level (2018 recommendation).	Pumps condition to be checked carefully with regards to sediment in hole potentially blocking impellers or sieves and lowering yield. Determine if pump is abstracting recommended volume of 10 108.8 m3/month. Assess water level over time data.
EC-N13-019	Pumping a 3rd (1.1 L/s) of recommended 2018 rate (3.2 L/s), establish if the borehole is operating at programmed capacity.	Consult water level over time data (if available).
EC-N13-022	Currently borehole is not working as it is awaiting a smaller pump, new borequip, and telemetry. This upgrade of the borehole has been put on hold as the service provider is awaiting payment from DWS. The process is 70 % underway, however as the municipality has changed banks, DWS is clarifying payment to correct accounts. Not adding what it should be (172 m^3/day) to an already low abstracted wellfield.	Expedite pump installation before Christmas 2019.
EC-N13-156	Water level below critical water level (2018 recommendations).	Assess water level over time data (if available).
EC-N13-160	Pumping at a lower rate than the 24-hour abstraction period rate.	Establish if borehole may produce more water by assessing water level over time data.
Kleinveeboer	Telemetry not yet installed. Access to pump house not possible – lock jam.	Install telemetry

Fig. 4.9 Extracts from the Wellfield Assessment completed by GEOSS in 2019. (Source: GEOSS December 19)

Table 23: Borehole Details EC-N11-033		
Borehole Name:	EC-N11-033	
Pump Station nr:	Only a pump installed	
Latitude:	-32.20078°	
Longitude:	24.54526°	
WL (mbgl during Pump/Rest and date):	14.3 mbgl- Pumping - 9:28	
2018 Abstraction Recommendations		
Duty Cycle (hr/day):	-	
Abstraction Rate (L/s):	Not Known	
Abstraction Rate (m3/day):	-	
Pump installation depth (m):	-	
Pump Type:	-	
Motor Type/size:	-	
Monitoring Installations		
Sample tap (Y/N):	N	
Flow metre (Y/N):	N not working	
Flow metre reading (m3)	N	
Observation pipe (Y/N):	Y	
Telemetry (Y/N):	N	
EC-N11-033		
Comments:	This borehole was initially reported as not working while driving past. Upon inspection it was pumping. No security measures in place. No working flowmeter. No logger. This is similar to condition found previously 17 Sep 2019	
Concerns:	Borehole not secure or monitored.	
Recommendations:	Secure borehole and install telemetry.	

Fig. 4.10 Extracts from the Wellfield Assessment completed by GEOSS in 2019. (Source: GEOSS December 19)

through efforts of trusted local community leaders who relied on existing relationships and social capital to push community-municipality collaboration. In terms of the drought, the collaboration revealed that despite the need for immediate disaster response, informed planning and the formation of relevant policies and frameworks must be achieved to ensure limited water resources are managed and adapted in anticipation of future drought events. Through the CGR, collaboration led to the funding and sharing of vital research to achieve this end. The study proved that even short-lived public-private collaboration can wield positive results when addressing complex problems like that of water insecurity. Additionally, the study revealed that enhancing and using non-physical capital, like social capital, can be vital in addressing slow-onset and repeated disasters like drought. Thus, the collaborative governance approach in the context of a drought serves as an addition to social capital in disaster studies research which predominantly observes fast-onset disaster contexts. It is recommended that further research on collaborative governance in small towns of amalgamated municipalities be observed to determine the extent to which amalgamations affect communication and public participation. Revisiting the status of the Nqweba Dam Dredging Project would add significantly to the outcomes observed in this chapter.

References

Aldrich, D. P., & Meyer, M. A. (2015). Social capital and community resilience. *The American Behavioral Scientist, 59*(2), 254–269.

Ansell, C., & Gash, A. (2008). Collaborative governance in theory and practice. *Journal of Public Administration Research and Theory: J-Part, 18*(4), 543–571.

Bartkus, V. O., & Davis, J. H. (Eds.). (2009). *Social capital: Reaching out, reaching in.* Edward Elgar Publishing Limited United Kingdom.

Basson, G., Bosman, E., Vonkeman, J. (2022). Reservoir sedimentation mitigation measures to deal with a severe drought at Graaff-Reinet, South Africa. In *E3S Web of Conferences 36*:03012 EDP Sciences.

Bendimerad, F. (2003). Disaster risk reduction and sustainable development. *World Bank Seminar on the role of local governments in reducing the risk of disasters, held in Istanbul Turkey, 28*, 57–75.

Besser, T. L., & Miller, N. J. (2015). Latino immigration, social capital, and business civic engagement in rural prairie towns. In J. M. Halstead & S. C. Deller (Eds.), *Social capital at the community level: An applied interdisciplinary perspective* (pp. 14–30). Routledge.

Bhandari, H., & Yasunobu, K. (2009). What is social capital? A comprehensive review of the concept. *Asian Journal of Social Science, 37*(3), 480–510.

Bourdieu, P. (1986). The forms of social capital. In J. Richardson (Ed.), *Handbook of Theory and Research for the Sociology of Education* (pp. 241–258). Greenwood Westport.

Busayo, E. T., & Kalumba, A. M. (2020). Coastal climate change adaptation and disaster risk reduction: A review of policy, programme, and practice for sustainable planning outcomes. *Sustainability, 12*, 6450.

Chamlee-Wright, E., & Storr, V. H. (2011). Social capital as collective narratives and post-disaster community recovery. *The Sociological Review, 59*(2), 266–282.

Claridge, T. (2018). Functions of social capital – bonding, bridging, linking. *Social Capital Research, 20*, 1–7.

Coffé, H., & Geys, B. (2007). Toward an empirical characterisation of bridging and bonding social capital. *Non-profit and Voluntary Sector Quarterly, 36*(1), 121–139.
Coleman, J. S. (1988). Social capital in the creation of human capital. *American Journal of Sociology, 94*, 95–120.
Cretney, R. M. (2016). Local responses to disaster: The value of community led post disaster response action in a resilience framework. *Disaster Prevention and Management, 25*(1), 27–40.
Dr Beyers Naude Local Municipality. (2021). *2017–2022 integrated development plan*. 5th Edition.
Emerson, K., & Nabatchi, T. (2015). Evaluating the productivity of collaborative governance regimes: a performative matrix. *Public Performance & Management Review, 38*(4), 717–747.
Emerson, K., Nabatchi, T., & Balogh, S. (2012). An integrative framework for collaborative governance. *Journal of Public Administration Research and Theory, 22*(1), 1–29.
Emery, M., & Flora, C. (2006). Spiralling-up: Mapping community transformation with Community Capitals Framework. *Community Development, 37*(1), 19–35.
Faling, W., Tempelhoff, J. W. N., & van Niekerk, D. (2012). Rhetoric or action: Are South African municipalities planning for climate change? *Development Southern Africa, 29*(2), 241–257.
Grootaert, C., & Van Bastelaer, T. (2002). *Understanding and measuring social capital: A multidisciplinary tool for practitioners*. The International Bank for Reconstruction and Development.
Hawkins, R. L., & Maurer, K. (2010). Bonding, bridging, and linking: How social capital operated in New Orleans following Hurricane Katrina. *British Journal of Social Work, 40*, 1777–1793.
Ma, Y., Lan, J., Thornton, T., Mangalagiu, D., & Zhu, D. (2018). Challenges of collaborative governance in the sharing economy: The case of free-floating bike sharing in Shanghai. *Journal of Cleaner Production, 197*, 256–365.
Mahlalela, P. T., Blamey, R. C., Hart, N. C. G., & Reason, C. J. C. (2020). Drought in the Eastern Cape regions of South Africa and trends in rainfall characteristics. *Climate Dynamics, 55*, 2743–2759.
Makaya, E., Rohse, M., Day, R., Vogel, C., Mehta, L., McEwen, L., Rangecroft, S., & Van Loon, A. F. (2020). Water governance challenges in rural South Africa: exploring institutional coordination in drought management. *Water Policy, 22*, 519–550.
Oh, Y., & Bush, C. B. (2016). Exploring the role of dynamic social capital in collaborative governance. *Administration & Society, 48*(2), 216–236.
Putnam, R. (1995). Bowling alone: America's declining social capital. *Journal of Democracy, 6*, 65–78.
Putnam, R. (2000). Bowling Alone. New York: Simon & Schuster Paperbacks.
Rogers, S. H., & Jarema, P. M. (2015). A brief history of social capital research. In J. M. Halstead & S. C. Deller (Eds.), *Social capital at the community level: An applied interdisciplinary perspective* (pp. 14–30). Routledge.
Satterthwaite, D. (2011). Why is community action needed for disaster risk reduction and climate change adaptation? *Environment & Urbanization, 23*(2), 339–349.
Smith, P. (2020). Water found at empty Nqweba Dam. Graaff-Reinet Advertiser. [Online] 4 February. Available at: https://www.graaffreinetadvertiser.com/News/Article/General/water-found-at-empty-nqweba-dam-202002040747 [Accessed 14 Aug. 2021].
Spocter, M. (2011). *Using geospatial data analysis and qualitative economic intelligence to inform local economic development in small towns: A case study of Graaff-Reinet, South Africa*. Nova Science Publishers.
Szreter, S., & Woolcock, M. (2004). Health by association? Social capital, social theory, and the political economy of public health. *International Journal of Epidemiology, 33*, 650–667.
Tierney, K. (2012). Disaster governance: Social, political, and economic dimensions. *The Annual Review of Environment and Resources, 37*, 341–363.
Visser, W. P. (2015). Water contestations in the Little Karoo: Liaisons between the Calitzdorp irrigation board and the Cailitzdorp (Kannaland) Municipality, 1912–2013. *The Journal for Transdisciplinary Research in Southern Africa, 11*(3), 186–207.
Voets, J., Brandsen, T., Koliba, C., & Verschuere, B. (2021). Collaborative governance. In *Oxford research encyclopaedia of politics*. Oxford University Press.

Water Resource Management (2022, May/June). *Siltation of SA dams – Working towards a sustainable solution.* The Water Wheel, 33–35. https://www.wrc.org.za/wp-content/uploads/mdocs/WW%20May-June%202022_DAM%20SILTATION.pdf

Woolcock, M., & Narayan, D. (2000). Social capital: Implications for development theory, research, and policy. *The World Bank Research Observer, 15*(2), 225–224.

Interviews

Interviewee A. (2021). *Telephonic interview on 2 September 2021 about the history of the GEDF and the state of the GEDF-DBNLM collaboration.* Permission not granted for recording.

Interviewee A. (2022). *Interview on 17 June 2022 about the GEDF-DBNLM collaboration and impact.*

Interviewee B. (2022). *Interview on 14 July 2022 about the relationship between the GEDF and DBNLM and the impacts of the collaboration.*

Interviewee C. (2022). *Interview on 12 July 2022 about the GEDF and the GEDF-DBNLM collaboration.*

Additional Sources Provided to Author by the Graaff-Reinet Economic Development Forum

GEOSS. (2019, December 9). *Assessment of existing wellfield infrastructure and monitoring equipment for Graaff-Reinet, Eastern Cape.*

Graaff-Reinet Economic Development Forum. (2020, February 19). *Constitution of the Graaff-Reinet Economic Development Forum.*

Chapter 5
Surviving the Tides: A Study of Small-Scale Fisheries and Community Livelihoods in Selected Coastal Towns of the Western Cape, South Africa

Samantha Williams

5.1 Introduction

Almost three-quarters of the world's population lives within 50 kilometres of the sea (World Economic Forum, 2022). These areas are significant as coastal and marine resources offer humans various goods and services. The oceans provide livelihoods, a place of recreation, beauty, wonder and untapped scientific discovery leading to improved human health, new medications, foods and advanced technologies (Visbeck, 2018; WWF, 2016). The oceans' resources alone sustain approximately three billion livelihoods in various coastal locations, the vast majority in developing countries (UN, 2021). Coastal environments, therefore, provide a wide range of valuable services and resources and support the livelihoods of millions of people worldwide. Estimates for small-scale fisheries alone contribute more than half of the world's marine and inland fish catch, employ more than 90 per cent of the world's 35 million capture fishers and support another 84 million people employed in jobs associated with fish processing, distribution and marketing (FAO, 2022). As a result, traditional or small-scale fisheries often see many individuals and communities facing precarious and vulnerable living and working conditions (Béné, 2003; Pomeroy et al., 2001). Poverty also remains widespread for millions of fishers and communities in coastal areas, with fishing activities primarily contributing to livelihoods with few alternative sources of income, employment or food.

Many coastal towns and settlements rely on fisheries and marine resources as a crucial livelihood and economic activity source. Therefore, in some instances, these areas are unique places often associated with traditional livelihoods such as fishing and tourism. However, the process of development and growth in coastal

S. Williams (✉)
Department of Geography and Environmental Studies, Stellenbosch University, Stellenbosch, South Africa
e-mail: samanthawilliams@sun.ac.za

© The Author(s), under exclusive license to Springer Nature Switzerland AG 2023

R. Donaldson (ed.), *Socio-Spatial Small Town Dynamics in South Africa*, GeoJournal Library, https://doi.org/10.1007/978-3-031-37142-4_5

environments has brought about significant changes in how people live and work. While such change can bring about new opportunities, it can also create challenges for coastal communities, including social tensions and environmental degradation (Fabinyi, 2010). For example, a recent study by Ayilu et al. (2023) examined the effects of the industrialisation of small-scale fisheries in Ghana and how this resulted in significant disturbances to local fishing communities' economic and social structures. These disruptions have resulted in disconnection, reduced incomes, increased conflicts, and social exclusion, adversely impacting women's social identity. The authors contend that as a consequence, small-scale fishing in Ghana's coastal areas is at risk, posing a significant threat to their livelihoods as coastal fishing communities have limited means to adapt.

In South Africa, the establishment and development of coastal towns and settlements were influenced by the country's early colonialism periods, various dispossession processes, apartheid planning, and unequal access and distribution of coastal and marine resources. Coupled with this, discriminative practices in the country's coastal provinces deprived most South Africans of fully benefitting from the country's rich coastal and marine resources. Regarding regional development, Isaacs et al. (2022) argue that the planned establishment of coastal settlements was based on the common model of company-established fishing towns. Coastal South African towns that developed due to this model include Saldanha Bay, St Helena Bay, Lamberts Bay, Jeffreys Bay, Port Nolloth, Hondeklipbaai and others (Isaacs et al., 2022). The establishment of these towns was influenced by various factors, such as physical, historical, economic, social, and political factors, resulting in ongoing differentiation. Therefore, Lemon (1991) added that unique approaches would be required to address change and transformation in such locations. This change will also be influenced by the political and economic orders adopted and how much those established structures are amendable to such transformation.

South Africa boasts a 3000-kilometre coastline which has provided for coastal communities for millennia. In this regard, many fishery traditions, systems, settlements, towns, and industries have developed, which grew from all of these activities. Historically, the activities and livelihoods of coastal inhabitants who made a living from marine resources were neglected in fisheries discourses, policy and management (Sowman, 2006). During the country's apartheid era, the fisheries sector was dominated by a few white-owned enterprises, and access to resources for black[1] coastal fisher communities was predominantly through employment in the industrial sector or through recreational or resource use permits (Cardoso et al., 2005; Van Sittert, 1992). Despite these restrictions, Van Sittert (2003) notes that many individuals and communities along the coasts of South Africa were involved in subsistence and small-scale fishing activities for decades. After democratisation in South Africa, the development of new legislation saw formal recognition of this

[1] Black' includes people from black African, coloured and Indian decent in the South African context.

marginalised sector (Harris et al., 2002). However, despite this recognition, little information about the fishers, their communities, what they harvested, and their livelihoods was available.

Today, small-scale fishing activities are recognised as critically important in livelihoods and providing food security for thousands. In South Africa, approximately 28,338 fisher households are found in nearly 147 fishing communities. Isaacs et al. (2022) add that the economic value of the fishing sector is R36 billion per annum, and the industry employs 27,000 people. However, many marine species are listed as fully utilised and highly valued resources, including west coast rock lobster and abalone, are over-exploited (DAFF, 2014). The small-scale fisheries sector has been plagued with complex problems. These challenges have primarily been associated with governance, particularly formal access and rights allocation processes. Furthermore, limited attention has been paid to the fishery's historical, cultural, and traditional elements and the contemporary social, economic, and resource sustainability challenges confronting numerous coastal communities in South Africa (Williams, 2021). This is why Isaacs et al. (2022) note that small-scale fishers remain a marginalised and vulnerable group.

Coastal areas are sites of complex interrelationships between humans and the natural environment. This chapter aims to draw attention to this interconnectedness and the livelihood strategies and activities of small-scale fisheries in coastal communities of South Africa. Within this context, an overview is provided on the role of fisheries governance processes, especially post-apartheid reforms that are relevant to coastal livelihoods. This section is deliberate here as these reforms have recognised and called for greater access to coastal resources. Still, failure to implement policy reforms successfully has meant little change for coastal livelihoods and even threaten future livelihoods. While recognition of other developmental reforms, such as the Land reform program, is relevant with regard to some coastal regions, mainly marine protected areas (see Sunde, 2014; Fay, 2009), this will not be highlighted here given the scope of the chapter. Tracing and documenting specific reforms in the small-scale fisheries sector will be explored as they are relevant to coastal livelihoods. Fishing as a primary livelihood strategy is discussed. By exploring examples and referencing selected study sites on the West coast of South Africa, this chapter will demonstrate how historical processes and contemporary challenges have shaped livelihoods in these locations and how fishing as a defining characteristic of these locations and livelihoods are increasingly becoming more complex. This chapter's results and discussion section also draw on the researcher's longstanding research engagements and ongoing interactions with community members at the study sites. Through these frequent interactions, the researcher and community members often discuss various issues related to fisheries management, community, and livelihood activities. By presenting these livelihood examples, it will highlight how different structural processes, which include social, economic and governance factors, have impacted and shaped these livelihoods.

5.2 Defining Concepts: Small-Scale Fisheries and Sustainable Livelihoods

Traditional or small-scale fishery usually evokes a mental image of a fisher setting out on a rowing boat into the vast unknown of the ocean. This image might hold partly true, but significant acknowledgement has been made of small-scale fishery livelihoods' varied and diverse nature (Short et al., 2021; Weeratunge et al., 2014). While the term 'traditional fishing community' has often been used to imply a homogenous group or romanticised notion of a coherent entity, these communities, like any other, have their social dynamics and divisions (Jentoft, 2020; Petersen, 2008). However, the term 'fishing community' is used internationally and locally and has a cultural dimension (Petersen, 2008). The term 'small-scale fisher community' is applied and encompasses local resource users who are distinguished from their large-scale commercial counterparts in terms of their fishing activities being less capital intensive, requiring low inputs of technology and labour as well and carried out in combination with other livelihood activities (Sowman, 2011).

International bodies such as the Food and Agricultural Organisation (FAO) provide a more expansive definition of small-scale and artisanal fisheries. Activities important for small-scale fisheries are found along the value chain, including pre-harvest, harvest and post-harvest activities undertaken by men and women. These activities are important in food security and nutrition, poverty eradication, equitable development and sustainable resource utilisation (FAO, 2015). On a global scale, this sector employs more than 90 per cent of the world's capture fishers and fish workers, about half of whom are women. This is, therefore, significant in terms of employment as these individuals are either full- or part-time fishers and fish workers or engaged in seasonal or occasional fishing. Some of these activities may even be recurrent sideline activities that may become significant in difficulties or when household or community stresses are experienced. As a diverse and dynamic sector, the specific subsector is often characterised by locational and seasonality factors. However, small-scale fisheries tend to be firmly anchored in local community structures, often reflecting longstanding historical links to adjacent fishery resources and have seen the evolvement of traditions and values specific to the local context. Therefore, fisheries represent a way of life for many small-scale fishers and fish workers, and their communities depend on access to fishery resources and land (FAO, 2015). Having access and secure tenure rights to land in the coastal or marine area are, in some cases, critical for ensuring and facilitating access to the fishery, accessory activities (including processing and marketing), housing and other livelihood support, as well as to contribute to overall wellbeing (Coulthard et al., 2011).

Despite limited recognition in the fisheries sector, small-scale fishing communities in South Africa still retain a degree of character and identity based on their livelihood activities. Therefore, the term 'small-scale fisher community' is relevant and acknowledged in contemporary discourses and legislation that underpins the sector. In the Small-Scale Fisheries Policy (2012), a small-scale fisher community "Means an established socio-cultural group of persons who are, or historically have been, fishermen and -women, including ancillary workers and their families; have shared aspirations and historical interests or rights in the harvesting, catching or processing of marine living resources; have a history of shared Small Scale fishing activity but, because of forced removals, are not necessarily tied to particular waters or geographic area; and were or still are operating near or in the seashore or coastal waters where they previously enjoyed access to marine living resources, or continue to exercise their rights communally in terms of an agreement, custom or law; and who regard themselves as a community" (South African Government, 2012:7).

No discussion on livelihoods is complete without recognising the seminal work of Chambers (1983) and later further developed in Chambers and Conway's (1992) "Sustainable rural livelihoods: practical concepts for the twenty-first century". It would be useful to clarify how livelihoods are understood here. Livelihoods comprise various activities, assets, relationships and mechanisms people (individuals and groups) employ to make a living and enable food security (Allison & Ellis, 2001). Chambers and Conway (1992) note that livelihoods include material and social assets and capabilities mediated through economic, political and institutional contexts. A key aspect of livelihoods recognises that they are diverse and complex (Carter & May, 1999). A sustainable livelihood approach (SLA) – adopted in a range of fields as a tool to understand livelihood assets, strategies and outcomes of poor, resource-dependent communities – recognises that people (primarily those dependent on natural resources) utilise a range of strategies to create an income and secure wellbeing (Allison & Ellis, 2001; Chambers, 1997). This approach aims "to identify what the poor have rather than what they do not have and to strengthen people's inventive solutions, rather than block or undermine them" (Allison & Ellis, 2001:378). This is exemplified in some fisher communities, for instance, where fishing activities are combined with small-scale agriculture or livestock farming and seasonal work if the opportunity exists. Scoones (1998) identifies three livelihood strategies in his conceptual analyses of rural poverty. These categories include (1) livelihood diversification, (2) agricultural intensification and extensification and (3) migration that households could draw on to cope and survive. With economic, political and social crises impacting livelihoods, many have been compelled to diversify income sources and strategies to contribute to their livelihoods (Brugère et al., 2008). Diversification strategies in coastal and fishery livelihoods include working as labourers, keeping livestock or poultry farming, informal trading and support from government social relief, which could generate additional income and reduce vulnerability (Brugère et al., 2008). While a livelihood in its simplest form is a means of sustaining a living, the SLA provides an operationalising mechanism for understanding this integrative concept. Key here is that it draws attention to capabilities, sustainability and equity as essential and which form the core of livelihood strategies (Chambers & Conway, 1992).

5.3 A Brief Overview of Historical Processes, Contemporary Small-Scale Fisheries Governance and Its Implications for Livelihoods

To understand the development and growth of small-scale fisheries livelihoods in South Africa, it is necessary to provide a brief historical account of processes that shaped and contributed to coastal areas. This section is followed by some background on the policy environment and post-apartheid reforms relevant to fisheries in South Africa. This brief overview of fisheries transformation (after apartheid) is highlighted here as it relates to broader political and developmental changes that occurred in the country and how this impacted the fisheries sector of South Africa. The outline here presents critical moments in the sector and events that have affected fisheries governance and livelihoods.

Various policies and legislation of the colonial and apartheid regimes, including the Black Land Act 27 of 1913, the Development Trust and Land Act 18 of 1936, the Group Areas Act 41 of 1951, the Coloured Labour Preference Policy (1955) and the 'Homelands' Policy, all effectively denied the majority of black South African citizens access to and ownership of vast stretches of South Africa's coastline and resources, and forced them to reside in designated areas in cities and the homelands (Hauck & Sowman, 2001). These discriminative policy interventions shaped settlement patterns across the country. The Coloured Labour Preference Policy (1955), for example, had a particular impact on the Western Cape province with its objective of coloured labour preference (Lemon, 2021). In this regard, it sought to exclude black 'Africans' from living in the province (Cardoso et al., 2005). As a result, many coastal towns and settlements along the West Coast of South Africa are inhabited by 'coloured'[1] communities. There exists a dearth of information and research on this topic. Still, according to Van Sittert (2003), before democratisation and during the colonial period, both black and coloured 'informal' fishers were engaged in the harvesting of a variety of coastal resources. However, due to discriminatory legislation introduced and the promotion of white ownership over coastal and fisheries resources, black South Africans had little to no legal access to these resources. Towards the end of apartheid, it became possible for these excluded groups to gain legal access by acquiring recreational or commercial access rights. However, in many instances, they were still denied access to the resources that they traditionally harvested because their harvesting methods and the quantities harvested did not conform to the conditions of the recreational or commercial fishing permits (Williams, 2013; Sowman, 2011).

In many of the post-1994 policy reforms and development planning, emphasis was placed on balancing fair and equitable access to natural resources and ensuring the sustainability of the country's resource base. In other words, the main objectives of these reforms were to 'right the wrongs of the past'. In this regard, section 25 (4) of the country's Constitution refers to access to natural resources, which is also based on ensuring equity. Reform in the fisheries sector of South Africa after the advent of democracy has witnessed many changes. The introduction and implementation of a fisheries policy in 1998, the Marine Living Resources Act (108 of 1998) (Amended as MLRA 5 of 2014), was only the beginning of a long reform process that is continuing. At the start of these reform processes, an established and thriving commercial fisheries sector existed, with its distribution of wealth, resources and power concentrated in favour of a small minority of white-owned companies (Hersoug & Holm, 2000). Isaacs and Gervasio (2012) note that, unlike the rest of Africa, Asia and Latin America, South Africa developed a robust commercial industry long before it developed an artisanal/ small-scale commercial sector or considered the role of these fishers in the commercial sector.

With the advent of democracy, the newly elected government had stated goals such as ensuring "the upliftment of impoverished coastal communities through improved access to marine resources and the sustainable management of those resources through appropriate strategies (ANC, 1994: 104). According to Isaacs (2006), this created enormous expectations among many marginalised fishing communities. As with many other sectors of the South African economy, the predominant issue in the fisheries sector that needed redress was access and ownership of fisheries resources. As highlighted above, the MLRA (1998b) was one of the post-apartheid policies relevant to the fishing industry and advocated three overarching goals: sustainable use of resources, equity and stability in the fishing industry. Other significant pieces of legislation that was relevant for coastal livelihoods and resources and that recognised and called for the protection of historically marginalised coastal and fishing communities included:

- The South African Constitution (1996) which guarantees socio-economic rights, including the right to access sufficient food;
- National Environmental Management Act (1998a:2), which "provides that the State must respect, protect, promote and fulfil the social, economic and environmental rights of everyone and strive to meet the basic needs of previously disadvantaged communities";
- Integrated Coastal Management Act (2008:30) which identifies that the state is the trustee of coastal areas and "must ensure that coastal areas are used, managed, protected, conserved and enhanced in the interest of the whole community and should take whatever reasonable measures it considers necessary to conserve and protect coastal areas for the benefit of present and future generations".

While these legislative provisions were relevant to small-scale fisheries activities, no single policy spoke to the needs and values of this dynamic and complex sector. Research and experience in implementing the MLRA demonstrated that many small-scale fishers claimed they were excluded from gaining legal access to fisheries resources and forced to harvest resources 'illegally' (Hauck, 2008; Isaacs, 2004). These sentiments culminated in various grievances being lodged against the responsible minister and were documented at the Fisher Peoples Human Rights Hearings held in 2003. Here, fishers and coastal communities could highlight their grievances and reiterate to the government that their right to food security and livelihoods was threatened. At these hearings, other critical issues were claims about fisher identities and how recognition should be made to individuals and communities who identified as fishers (Williams, 2013).

The challenges and failures to address the needs of fishers and coastal communities culminated in the initiation of a court case in 2004 by fishers who believed that they had a legitimate right to access marine resources. Supported by the Legal Resources Centre (LRC), the landmark case of Kenneth George and others versus the Minister of Environmental Affairs and Tourism (DEAT) focussed on fishers' rights to secure a livelihood. It was settled in an out-of-court agreement in 2007. The judge signed a court order stating that all parties concerned would prepare a new legislative and policy framework that would include all traditional and small-scale fishers in South Africa (Sunde et al., 2011). It was agreed by the then minister to embark on a participatory policy process and put in place an interim measure which saw the awarding of what is known as 'Interim Relief Permits' (IRPs) to fishers. This was done using the recreational Total Allowable Catch (TAC) allocation. The minister was allowed to deviate from the standard regulations (i.e. allowing fishers access to the sea and to catch a maximum of 20 crayfish on any given day of the week and a combination of 30 specified line fish per day) (MDT, 2008). This catch could be sold, which is not generally allowed under recreational fishing regulations. It was furthermore stipulated that only a maximum of 1000 fishers who met stringent criteria could benefit from this concession. Therefore, IRPs were issued to individuals identified by their fishing communities as *bona fide* fishers, and measures to control and evaluate the process were also instituted. It was envisaged that the IRPs would be awarded to fishers until the Small-Scale Fisheries Policy (SSFP) was implemented. The SSFP was finalised in 2012. The policy's overarching aims are to provide redress and recognition to the rights of small-scale fisher communities in South Africa who were previously marginalised and discriminated against in terms of racially exclusionary laws and policies, individualised permit-based systems of resource allocation and insensitive impositions of conservation-driven regulation (South African Government, 2012). In terms of supporting and developing livelihoods, the policy aims to facilitate the development of a multi-sectoral and integrated approach that incorporates the following:

(a) economic ssgrowth and transformation of the economy to create decent work and sustainable livelihoods;
(b) vibrant, equitable, sustainable rural communities contributing towards food security for all; and
(c) protect and enhance our environmental assets and natural resources through sustainable resource management and use.

Various aspects of policy implementation are lagging. This has been a factor which has led to observed discontent and frustration in the sector. In this regard, many individual fishers and communities have insecure rights, contributing to a precarious livelihood.

5.4 Livelihood Strategies of Small-Scale Fishing Communities in South Africa: The Case of Lamberts Bay and Ebenhaeser, West Coast

The following section introduces the communities of Lamberts Bay and Ebenhaeser, where fishing activities play a crucial role in supporting the livelihoods of local fishers. In both these locations, long histories and involvement in fisheries activities have been documented, and these practices are central to livelihood activities in these communities.

5.4.1 Lamberts Bay

Lamberts Bay is a coastal town named after Admiral Lambert of the British Navy, who undertook a marine survey of the bay between 1826 and 1840 (Lamberts Bay Area plan 2017–2022, n.d.). It is noted that the town was used as a layup for British warships during the war of 1900–1902. Once known as the 'Crayfish mecca' of the West coast, the coastal town of Lamberts Bay has a long and rich history of fishing that contributed to the development and activities of the town. Therefore, its reason and existence are linked to the fishing industry, and the harbour is still operational today. Located +/– 290 kilometres northwest of Cape Town on the West Coast (See Fig. 5.1), this town was the founding location for one of South Africa's biggest fishing operations, Oceana. While the group has an expansive network of fish processing and production activities across southern Africa, its operations started in 1918 in Lamberts Bay. They were initially known as the Lamberts Bay Canning company limited (Oceana Group, 2022). Historically canned crayfish (or *kreef* as it is locally known) was processed at its former operation at Lamberts Bay (See Fig. 5.2) and exported to European markets. By 1927, 500 people wholly depended on the company for their livelihoods. With technological advances and growth, the canning company merged with other smaller rock lobster processing companies in Cape

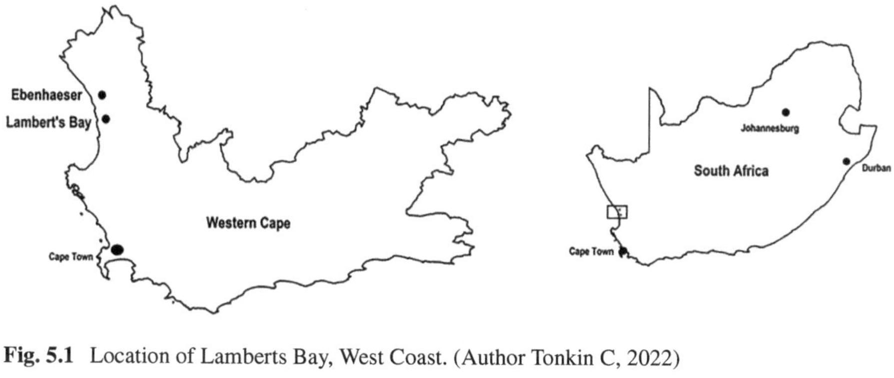

Fig. 5.1 Location of Lamberts Bay, West Coast. (Author Tonkin C, 2022)

Fig. 5.2 A fisher on his boat in the Lamberts Bay harbour. The white building in the background is the former crayfish factory. (Source: Author)

Town and Hout Bay. Their pooled quotas resulted in South African Sea Products holding the largest rock lobster quota in the country (Oceana Group, 2022). During its development and active operation in Lamberts Bay in earlier years, this group enjoyed and participated in the South African fishing industry, with towns like Lamberts Bay's development (and many livelihoods) historically dependent on these activities.

Lamberts Bay is part of the Cedarberg local municipality, forming part of the West Coast District municipality. The 2011 national census recorded a population of just over 6100 people. As a coastal town on the West coast, it has historically been described as the 'Crayfish mecca' and the 'Diamond of the West coast' due to its pristine white beaches and wildlife (Lamberts Bay Area plan 2017–2022, n.d.). The area's wildflowers and the Bird Island Nature Reserve are major tourist attractions, the nesting ground of Cape gannets, penguins and other bird species. The processing factory established for the packaging of crayfish has changed operation. It is now a potato chips facility that still contributes to the growth and economy of the town (Lamberts Bay Area plan 2017–2022, n.d.).

The establishment and activities of fishing companies like Oceana in coastal towns are what Isaacs et al. (2022) referred to as company-established fishing towns. Sea Harvest is another major fishing company and a common household name in South Africa. Its development and activities are inextricably linked to the growth of the town of Saldanha Bay (approximately 110 kilometres northwest of Cape Town). Welman and Ferreira (2017) trace these developments and the hake deep trawling industry's role in the Greater Saldanha Bay area. As one of the two major role players in the deep trawling industry in South Africa, Sea Harvest was established in 1964 in Saldanha Bay and continues to dominate the fishing industry in the area. The company is the largest employer in Saldanha Bay, and in the greater West coast area, over 15% of household income is derived from the company (Sea Harvest Group, 2022). For the town of Saldanha, this number is at 30% (Sea Harvest Group, 2022). The company is the major employer in the Greater Saldanha Bay region, with an estimated 2121 permanent and 360 contracted staff members (Welman & Ferreira, 2017). The number of employed individuals supports an average of five other family members (more than 12,000 people), which translates to between 25 and 30% of all household income in the Greater Saldanha Bay municipal area being generated through Sea Harvest's activities (Groenewald, 2015 quoted in Welman & Ferreira, 2017).

Livelihood activities which include fishing, remain important to households in Lamberts Bay. Two recent studies documented the socio-economic conditions of households where fishing remains a primary income source. Nthane's (2015) study aimed to establish a profile of the Lamberts Bay fishers and understand their perceptions of the Small-scale fisheries policy. This investigation aimed to understand the fishers' socio-economic contexts and determine if socio-economic differences existed between the fisher groups. The study's participants included West coast rock lobster (WCRL) rights holders and fishers who were awarded IRPs. What was found was that even though both groups showed almost identical historical dependence on fishing, differentiated rights allocations created unequal income opportunities and divisions amongst fishers (Nthane, 2015).

Notwithstanding these challenges, the value and contribution of fishing activities were still the primary income sources. A limited number of fishers (WCRL permit holders) included in Nthane's study indicated that they also work as WCRL trap makers. Wives of fishers employed outside the fisheries sector either worked in the potato chips factory, in the service industry or as domestics (Nthane, 2015).

Similarly, Jefthas (2019) conducted a household survey with twenty respondents in Lamberts Bay and found that eighteen participants indicated that their primary income is derived from fishing activities. From this sample, all primary harvesters were male fishers (n = 16), and post-harvest workers included three women and one male. The income derived from selling fisheries resources and activities ranged between R500-R2500 (Jefthas, 2019). Most incomes derived were spent on food or paying for household necessities. With limited employment opportunities, individuals would engage in ad hoc work, seasonal employment, and income from social grants as a supplement provides a lifeline for many households (Auld & Feris, 2022). While exploring perceptions of alternative sources or opportunities to secure a livelihood, Jefthas (2019) found that fishers did not believe such opportunities existed in Lamberts Bay. All participants concluded that there "are no other jobs available for them to earn a living, and because Lamberts Bay is a fishing town participating in fishing activities is the only way they can generate an income" (Jefthas, 2019: 34). Similar sentiments expressed by fishers have been shared with the author. However, fishers have also noted that fishing has become an increasingly expensive livelihood activity. For instance, fishermen who operate boats and work with crew members have highlighted that expenses such as fuel and bait have rendered it nearly impossible to profit from fishing activities (Informal discussion with fishermen in Lamberts Bay, 18 August 2022). Since there is no guarantee of a successful catch, fishermen must evaluate the increasing supply costs and how the crew will divide the income. These considerations and factors also impact their decisions to engage in fishing (Informal discussion with fishermen in Lamberts Bay, 18

August 2022). Fishers and other community members frequently mentioned the government's awareness of their challenges. However, despite almost three decades of democracy and policy reforms, the promised benefits have yet to materialise (Informal discussion with fishermen in Lamberts Bay, 18 August 2022).

5.4.2 *Ebenhaeser*

Further north of Lamberts Bay is a small rural settlement located on the banks of the Olifants River (See Fig. 5.1). Ebenhaeser forms part of the Matzikamma local municipality, which is located within the larger West Coast District municipality. Neighbouring coastal towns include Doringbay and the holiday town of Strandfontein. The 2011 census recorded a population of 1305 people in this community. For decades, traditional net fishers have made their living from harvesting *Liza richadsonni* (southern mullet) or *harders* from the Olifantsriver estuary (Williams, 2021; Sowman, 2010). These practices continue today (See Fig. 5.3) with fishers harvesting mullet from the Olifantsriver estuary and participating in fishing activities in other coastal towns.[2] Apart from seasonal work on neighbouring farms in the area, few alternative employment opportunities are available to fishers and community members (Informal discussion with fishers at Ebenhaeser, 15 August 2022). Government social support grants are also important to supplement household incomes. While gillnet fishing activities have been prohibited in many South African estuaries, this practice continues here with harvesting and access restrictions being in place.

In this fishery, gillnets and small rowing boats are the primary tools for engaging in fishing activities. No capital is needed to purchase bait or fuel for boat engines. Fishers sell their catches to communities in neighbouring towns, including Lutzville and Doringbay, and as far afield as Lamberts Bay (Informal discussion with fisher at Ebenhaeser, 15 August 2022). Fetching high prices for *harders* are not often considered and fresh fish could be sold for as little as R2.00- R2.50[3] per fish. No cold storage facilities are available in the Ebenhaeser community. Fresh fish must be sold almost immediately after landing (Informal discussion with fishers at Ebenhaeser, 15 August 2022). Fishers have indicated that with cold storage and proper processing facilities, they might be able to extend their product offering to other markets. However, while this is absent, they depend on existing markets to buy their catches. Where surplus catches are not sold fresh, this is salted and dried and sold as '*bokkoms*' (See Fig. 5.4), which are regarded as a local delicacy of the West coast.

The national fisheries authorities regulate the activities of the net fishers of the area. This means that only 45 rights allocations are legally valid for the Oliftantsriver

[2] A number of fishers from the Ebenhaeser community work with fishers from Doringbay and Lamberts Bay fishing along the coast, especially during specific seasonal catch times.

[3] Approximately $0.15 based on a ZAR/ US dollar exchange of R1/ $0.55.

Fig. 5.3 Fishers set out on a fishing trip at the Olifantsriver estuary. (Source: Author)

Fig. 5.4 Dried 'bokkoms' at Ebenhaeser. (Source: Author)

gillnet fishery, and each rights holder is allowed one crew member. The permit awarded to the fishers is only for harvesting southern mullet. The continued practice of gillnet fishing at the Olifantsriver estuary has been a contested issue in small-scale fisheries management. In 2005 a policy directive called for the phasing out of gillnet activities at the site and for the community to explore alternative livelihood options. This policy directive was considered unjust to community members as they were neither informed about nor consulted on the matter. What followed this was years of negotiation and the involvement of litigation experts. What could be regarded as a success after the completion of these processes was the recognition by management authorities that fishing for this community extended well beyond income-generating opportunities (Williams, 2021). Several studies have documented this community's livelihood and socio-economic conditions (see Williams, 2013, 2021; Hushlak, 2012; Sowman, 2010). These have all shown that fishing is an important livelihood activity and critical for food security. Furthermore, this practice for the community is inextricably linked to processes of identity formation and a sense of place (Williams, 2021). In this regard,fishers and community members have often expressed that they are born fishermen and women and that their activities form part of their individual and community identity.

In Nthane's (2015) and Jefthas's (2019) studies in Lamberts Bay, the authors also emphasised cultural aspects of fishing practices as critical to fishers and their communities. The historical, emotional attachments and personal relationships that fishers have with the ocean underpin their motivations for engaging in these activities and have been recorded by numerous studies (Jentoft, 2020; Young et al., 2016; Pollnac & Poggie, 2008). In this sense, these varied motivations lie behind why fishers do what they do.

While the traditional fishing practices of the Olifantsriver fishers have not gone uncontested, another threat is likely further to impact livelihoods in this area. In recent years, increasing interest in mining activities and prospecting along the West coast of South Africa have been linked to discourses about how the 'Blue economy' models can potentially enhance coastal livelihoods. While the term 'Blue economy' has no agreed definitions, it has been used as a synonym for the term 'ocean economy' (Loureiro et al., 2022). This would therefore refer to any economic activity in the maritime sector, whether this is defined as sustainable (and aligned with green economy principles) or not.

Aspects associated with the 'Blue economy' have raised questions and concerns for the livelihoods of coastal communities. Multinational oil companies like Shell have been implicated in seismic activities off South Africa's West coast, and in 2021, coastal communities challenged the company in court. In March 2022, coastal communities won their battle when a judge ruled in their favour and condemned seismic surveys for oil exploration (Isaacs et al., 2022; Yeld, 2022). However, following this ruling, another multinational company, Total, applied for oil exploration on the West coast.

For the community of Ebenhaeser, this has been a key concern as these activities could impact ecosystem health and, consequently, the resources they harvest (Informal discussion at Ebenhaeser with fishers, 15 August 2022). Fishers and community members are working closely with non-governmental organisations (NGOs) and academics to monitor potential mining activities. Regarding the Olifantsriver estuary, Sowman (quoted in Christianson, 2021) notes that the number of environmental authorisations granted for mining is of concern. An example that illustrates this is the appeal made against environmental authorisation for prospecting on coastal land at the Olifantsriver estuary's northern banks, which is one of the important estuaries in South Africa (Christianson, 2021). A concern highlighted is that if prospective activities yield good results, the next step would be mining (Sowman quoted in Christianson, 2021). The fishing community is concerned about these activities, amongst others, and how machinery (used to prospect) may deter the mullet from entering the river mouth, jeopardising their livelihoods and threatening their food security (Christianson, 2021). The decision to allow prospecting is difficult to understand because a proposal has been made to declare a community-protected area at the Olifantsriver estuary (relevant stakeholders have approved the plan). There have been delays in finalising the process. The community hopes that once a community-protected area is declared, it will give them greater decision-making power to exercise and comment on proposed mining activities in the area (Informal conversations with fishers at Ebenhaeser, 15 August 2022).

The onslaught of such attempts raises concerns for coastal areas and Isaacs et al. (2022) question how long coastal communities could hold off big oil and gas companies from their fishing grounds (Isaacs et al., 2022).

5.5 Concluding Thoughts

Policy reform in the small-scale fisheries sector has undoubtedly provided recognition to thousands of individuals who make their livelihoods from South Africa's coastal and fisheries resources. However, Isaacs et al. (2022) add that this recognition is insufficient in addressing poverty and creating sustainable livelihoods for coastal communities. "Most small-scale fishing communities in the four maritime provinces are in rural towns and villages where unemployment is higher than in urban areas. Thus, in these areas, fishing provides one of the most potent or only avenues for livelihoods and income" (Isaacs et al., 2022:10).

Poverty is a South African reality. Concerns over resource constraints and availability have been acknowledged, but these constraints can undermine food security and poverty alleviation efforts. In many coastal areas, further constraints such as lack of access to capital, technology and limited alternative employment mean that coastal livelihoods remain locked in a vicious poverty cycle. Even more concerning are the constraints and challenges associated with policy issues and how access rights to fisheries resources remain contested in fisheries governance.

South Africa's coastal areas are considered a national asset and crucial for the country's economy and the wellbeing of coastal inhabitants (Peter, 2021). In this regard, coastal towns and settlements are important centres of livelihoods and economic activities for millions of people. Growth and development activities must prioritise social, economic and environmental considerations to address some of the challenges faced in coastal areas. This will require coordination and collaboration between stakeholders, including local communities, government and private sector actors.

With limited or no prospects for livelihood diversification, small-scale fishing communities in South Africa will remain exposed to vulnerabilities, have insecure livelihood activities and social inequality will persist. Policies such as the SSFP advocate for co-management arrangements that will support coastal livelihoods. Still, considering the relationship between current livelihood strategies, management approaches and feasible alternative livelihood opportunities needs exploration. This will go a long way in improving efforts to reduce vulnerability in coastal towns and settlements whilst trying to maximise food security and poverty reduction strategies. With the importance of small-scale fisheries being recognised in South Africa, current management approaches need to demonstrate the commitment of various stakeholders to meet the objectives of the SSFP. Supporting small-scale fisheries' growth and sustained activities is about improving and enhancing livelihoods and who people are.

References

African National Congress (ANC). (1994). *Reconstruction and development programme: A policy framework*. South Africa.

Allison, E., & Ellis, F. (2001). The livelihoods approach and management of small-scale fisheries. *Marine Policy, 25*, 377–388.

Auld, K., & Feris, L. (2022). Addressing vulnerability and exclusion in the South African small-scale fisheries sector: Does the current regulatory framework measure up? *Maritime Studies, 21*, 533–552.

Ayilu, R., Fabinyi, M., Barclay, K., & Bawa, M. A. (2023). Blue economy: Industrialisation and coastal fishing livelihoods in Ghana. *Reviews in Fish Biology and Fisheries*. https://doi.org/10.1007/s11160-022-09749-0

Béné, C. (2003). When fishery rhymes with poverty: A first step beyond the old paradigm on poverty in small-scale fisheries. *World Development, 31*(6), 949–975.

Brugère, C., Holvoet, K., & Allison, E. (2008). *Livelihood diversification in coastal and inland fishing communities: Misconceptions, evidence and implications for fisheries management.* Working paper, sustainable Fisheries livelihoods Programme (SFLP). .

Cardoso, P., Fielding, P., & Sowman, M. (2005). *Overview and analysis of social, economic and fisheries information to promote artisanal fisheries management in the BCLME region – South Africa.* Final report and recommendations, Cape Town, South Africa: Environmental Evaluation Unit, University of Cape Town.

Carter, M., & May, J. (1999). Poverty, livelihoods and class in rural South Africa. *World Development, 27*(1), 1–20.

Chambers, R. (1983). *Rural development: Putting the last first*. Longman.

Chambers, R. (1997). Responsible wellbeing: A personal agenda for development. *World Development, 25*(11), 1743–1754.

Chambers, R., & Conway, G. (1992). Sustainable rural livelihoods: Practical concepts for the 21st century. *IDS discussion paper,* 296.

Christianson, B. (2021, December 27). Olifants River set to become a new home of mining — To the detriment of residents and the environment as a whole. *Daily Maverick*.

Coulthard, S., Johnson, D., & McGregor, A. (2011). Poverty, sustainability and human wellbeing: A social wellbeing approach to the global fisheries crisis. *Global Environmental Change, 21*, 453–463.

Department of Agriculture Forestry and Fisheries (DAFF). (2014). *Status of the south African marine fishery resources 2014*. DAFF, Republic of South Africa.

Fabinyi, M. (2010). The intensification of fishing and the rise of tourism: Competing coastal livelihoods in the Calamianes Islands, Philippines. *Human Ecology, 38*, 415–427. https://doi.org/10.1007/s10745-010-9329-z

FAO. (2015). *Voluntary guidelines for securing sustainable small-scale fisheries in the context of food security and poverty eradication*. Rome.

FAO. (2022). *The State of world fisheries and aquaculture 2022*. Towards Blue Transformation. Rome. https://doi.org/10.4060/cc0461en

Fay, D. (2009). Land tenure, land use, and land reform at Dwesa-Cwebe, South Africa: Local transformations and the limits of the State. *World Development, 37*(8), 1424–1433.

Harris, J., Sowman, M., Branch, G., et al. (2002). The process of developing a management system for subsistence fisheries in South Africa: Recognising and formalising a marginalised fishing sector in South Africa. *South African Journal of Marine Science, 24*, 405–424.

Hauck, M. (2008). *Rethinking small-scale Fisheries compliance: From criminal justice to social justice* [Doctoral thesis]. University of Cape Town.

Hauck, M., & Sowman, M. (2001). Coastal and fisheries co-management in South Africa: An overview and analysis. *Marine Policy, 25*, 173–185.

Hersoug, B., & Holm, P. (2000). Change without redistribution: An institutional perspective on South Africa's new fisheries policy. *Marine Policy, 24*, 221–231.

Hushlak, A. (2012). *Integrating traditional ecological knowledge in South Africa's small-scale fisheries: The Olifants Estuary Gillnet Fishery*. [Masters dissertation]. University of Cape Town.

Isaacs, M. (2004). *Understanding the social processes and the politics of implementing a New Fisheries Policy, the Marine Living Resource Act 18 of 1998, in South Africa*. [Doctoral thesis]. University of the Western Cape.

Isaacs, M. (2006). Small-scale fisheries reform: Expectations, hopes and dreams of a better life for all. *Marine Policy, 30*(1), 51–59.

Isaacs, M., & Gervasio, H. (2012). The dualistic nature of fisheries and policy responses in Mozambique and South Africa. In L. Masters & E. Kisiangani (Eds.), *Natural resources governance in southern Africa* (pp. 37–64). African Books Collective.

Isaacs, M., Hara, M. M., Dennis T. L., Rouhani, Q. A., Mannarino, C., Jaffer, N. (2022). *A situational analysis of small-scale fisheries in South Africa: From vulnerability to viability*. V2V Working Paper 2022-9. V2V Global Partnership, University of Waterloo, Canada.

Jefthas, T. (2019). *The Contribution of small-scale fisheries to livelihoods: A case study of Lamberts Bay*. [Honours thesis]. Stellenbosch University.

Jentoft, S. (2020). Life above water: Small-scale fisheries as a human experience. *Maritime Studies, 19*, 389–397.

Lamberts Bay Area Plan 2017–2022 (n.d.). http://cederbergmun.gov.za/sites/default/files/documents/Lambertsbay%20Area%20Plan.pdf

Lemon, A. (1991). The apartheid city. In A. Lemon (Ed.), *Homes apart: South Africa's segregated cities* (pp. 1–25). David Phillip.

Lemon, A. (2021). The apartheid city. In A. Lemon, R. Donaldson, & G. Visser (Eds.), *South African urban change three decades after apartheid: Homes still apart?* (pp. 1–16). Springer.

Loureiro, T. G., Du Plessis, N., & Findlay, K. (2022). Into the blue – The blue economy model in operation Phakisa 'unlocking the ocean economy' programme. *South African Journal of Science, 118*, 11–12.

MDT. (2008). *Small-scale Fisheries policy proposal developed at the Masifundise CWIC National Workshop*. Masifundise Development Trust (MDT), Cape Town.

Nthane, T. (2015). *Understanding the livelihoods of small-scale fishers in Lamberts Bay: Implications for the new Small-scale Fisheries Policy*. [Masters thesis]. University of Cape Town.

Oceana Group. (2022). *Our history*. https://oceana.co.za/about-oceana/our-history/#:~:text=It%20all%20started%20in%201918,Lambert's%20Bay%20Canning%20Company%20Ltd.&text=The%20factory%20began%20canning%20Cape,kreef'%20for%20export%20to%20France.

Peter, R. (2021). Integrated coastal management in South Africa: Achievements and challenges over 8 years of implementation. *Journal of Coastal Zone Management, (Short Communication), 24*(6), 1–1.

Petersen, C. (2008). *Contesting citizenship: Civil society struggles over livelihood and educational access in coastal settlements of the Western Cape, South Africa*. [Doctoral thesis]. University of Edinburgh.

Pollnac, R., & Poggie, J. (2008). Happiness, wellbeing and psychocultural adaptation to the stresses associated with marine fishing. *Human Ecology Review, 15*, 194–200.

Pomeroy, R., Katon, B., & Harkes, I. (2001). Conditions affecting the success of fisheries co-management: Lessons from Asia. *Marine Policy, 25*(3), 197–208.

Scoones, I. (1998). Sustainable rural livelihoods: A framework for analysis. *IDS Working Paper, 72*, 1–22.

Sea Harvest Group. (2022). *Our story*. https://www.seaharvest.co.za/our-story/

Short, R., Gelcich, S., Little, D., et al. (2021). Harnessing the diversity of small-scale actors is key to the future of aquatic food systems. *Nature Food, 2*, 733–741. https://doi.org/10.1038/s43016-021-00363-0

South African Government. (1996). *Constitution of the Republic of South Africa, Act 108 of 1996*. Pretoria, South Africa.

South African Government. (1998a). *National Environmental Management Act 107 of 1998*. Pretoria.

South African Government. (1998b). *Marine Living Resources Act 18 of 1998*. Pretoria.

South African Government. (2008). *National Environmental Management: Integrated Coastal Management Act 24 of 2008*. Pretoria, South Africa.

South African Government. (2012). *Policy for the small-scale fisheries sector in South Africa*. Department of Agriculture, Forestry and Fisheries, South Africa.

Sowman, M. (2006). Subsistence and small scale fisheries in South Africa: A ten year review. *Marine Policy, 26*, 60–73.

Sowman, M. (2010). An evolving partnership: Collaboration between university 'experts' and net-fishers. *Gateways: International Journal of Community Research and Engagement, 2*, 119–143.

Sowman, M. (2011). New perspectives in small scale fisheries management: Challenges and prospects for implementation in South Africa. *African Journal of Marine Science, 33*, 297–311.

Sunde, J. (2014). *Customary governance and expressions of living customary law at Dwesa-Cwebe: contributions to small-scale fisheries governance in South Africa*. [Doctoral thesis]. University of Cape Town.

Sunde, J., Sowman, M., Smith, H., & Wicomb, W. (2011). *Emerging proposals for governance of tenure in small – scale fisheries in South Africa*. Paper prepared for the Governance of Tenure for Responsible Capture Fisheries. A UN/FAO Fisheries and Aquaculture Department Initiative.

United Nations. (2021). UN Secretary-General António Guterres. *Press Statement*, SG/SM/207541 June 2021. https://press.un.org/en/2021/sgsm20754.doc.htm.

Van Sittert, L. (1992). *Labour, Capital and the State in the St Helena Bay Fisheries C*. 1856-C. 1956. [Doctoral thesis]. University of Cape Town.

Van Sittert, L. (2003). The tyranny of the past: Why local histories matter in the South African fisheries. *Ocean and Coastal Management, 46*(1–2), 199–219.

Visbeck, M. (2018). Ocean science research is key for a sustainable future. *Nature Communications, 9*(690), 1–4. https://doi.org/10.1038/s41467-018-03158-3

Weeratunge, N., Béné, C., Siriwardane-de Zoysa, R., Charles, A., Johnson, D., Allison, E., Nayak, P., & Badjeck, M.-C. (2014). Small-scale fisheries through the wellbeing lens. *Fish and Fisheries, 15*(2), 255–279.

Welman, L., & Ferreira, S. (2017). Sea Harvest: Back(fish)bone in Saldanha Bay's local and regional economy? *Local Economy, 32*(6), 487–504.

Williams, S. (2013). *Beyond rights: Developing a conceptual framework for understanding access to coastal resources at Ebenhaeser and Covie, Western Cape, South Africa*. [Doctoral thesis]. University of Cape Town.

Williams, S. (2021). I fish because I am a fisher: Exploring livelihood and fishing practices to justify claims for access to small-scale fisheries resources in South Africa. *Theological Studies, 77*(3), 1–9.

World Economic Forum. (2022). *Only 15% of the world's coastlines remain in their natural state*. https://www.weforum.org/agenda/2022/02/ecologically-intact-coastlines-rare-study/

WWF-SA. (2016). *Oceans facts and futures: Valuing South Africa's ocean economy*. WWF-SA.

Yeld, J. (2022, 21 June). West Coast breakthrough: environment activists and mining company reach agreement. *GroundUp*. https://www.groundup.org.za/article/environmental-activists-reach-agreement-with-australian-mining-company/

Young, M., Foale, S., & Bellwood, D. (2016). Why do fishers fish? A cross-cultural examination of the motivations for fishing. *Marine Policy, 66*, 114–123.

Chapter 6
The State and Implications of Housing Provision in Small Towns: Experiences in the Raymond Mhlaba Local Municipality, Eastern Cape Province

John Ntema, Robert Mongwe, and Eliot Mathebula

6.1 Introduction

Despite the country's undergoing what Hoogendoorn and Visser (2016:95) termed "a phase of rural depopulation and small town decline", the geography of the South African urban landscape is dominated by small towns, rather than cities and large towns (Nel & Hill, 1998; Ntema & Venter, 2016). There are about 500 small towns in South Africa (Van Niekerk & Marais, 2008; Centre for Development and Enterprise, 1996). The global significance of small towns is also undeniable. It is currently estimated that about 60% of the world's urban dwellers reside in small- and medium-sized urban areas, with small towns in sub-Saharan Africa being home to more than 196 million urban dwellers (Agergaard et al., 2018). However, despite their geographical dominance, there is currently a dearth of scientific knowledge and research on small towns, particularly in South Africa (Nel & Hill, 1998; White, 2004; Gibb & Nel, 2007; Ballard & Steyn, 2013; Hoogendoorn & Visser, 2016; Ntema, 2021). Missing in key themes covered in a few studies done on small towns and their rural local municipalities in South Africa is, amongst other things, a comprehensive review of low-income housing provision and its possible role in spatial redress (see also Hoogendoorn & Visser, 2016). It is in light of this knowledge gap in the existing literature that we found it appropriate and relevant to undertake this study on small towns and low-income housing provision in the rural Raymond Mhlaba local municipality, Eastern Cape province.

According to Xusa (2005), there are over 70 small towns in the Eastern Cape province. As widely confirmed in the existing literature, small towns are usually

J. Ntema (✉) · E. Mathebula
Department of Development Studies, University of South Africa, Pretoria, South Africa
e-mail: lejone12345@gmail.com

R. Mongwe
Department of Human Settlement, University of Fort Hare, Alice, South Africa

© The Author(s), under exclusive license to Springer Nature Switzerland AG 2023

R. Donaldson (ed.), *Socio-Spatial Small Town Dynamics in South Africa*, GeoJournal Library, https://doi.org/10.1007/978-3-031-37142-4_6

synonymous with poor, rural local municipalities. As one of the former Bantustans and the poorest rural provinces (see also Pauw, 2005; Hoogendoorn & Visser, 2016; Stats SA, 2021), the Eastern Cape is home not only to most small rural towns but to most rural local municipalities in the country as well (Bank et al., 2010; Westaway, 2012; Ngcayi, 2013). One such rural local municipality is the Raymond Mhlaba local municipality located in the predominantly rural Amathole District Municipality. As shown in Fig. 6.1, the small towns that comprise the Raymond Mhlaba local municipality include Alice, Fort Beaufort, Bedford, Adelaide, Middledrift, Hogsback, and Seymour (Raymond Mhlaba Local Municipality, 2018a). Like elsewhere in other small towns and their rural local municipalities in the country, particularly in the Eastern Cape province, the urban landscape of the Raymond Mhlaba local municipality and Amathole District by extension, bears some remnants of the inherent apartheid housing landscape. Amongst other things, this urban landscape is characterised by multiple socio-spatial inequalities, chiefly a housing backlog and peripheral location of the state-funded, low-income housing developments (Raymond Mhlaba Local Municipality, 2018a; Ntema, 2021). Notwithstanding the strides made since the dawn of democracy in 1994, the current housing landscape in the Raymond Mhlaba local municipality presents an interesting lens through which a critical review of government performance in particular, could be done. Against the background above, the chapter contends that despite national and provincial interventions, including various localised interventions and strategic planning documents, such as the Integrated Development Plan (IDP), the Spatial Development Framework (SDF) and the Small Town Regeneration Development Strategy by the Raymond Mhlaba local municipality, all the seven, small towns in this rural local municipality continue to struggle with a threefold housing related challenge: first, a

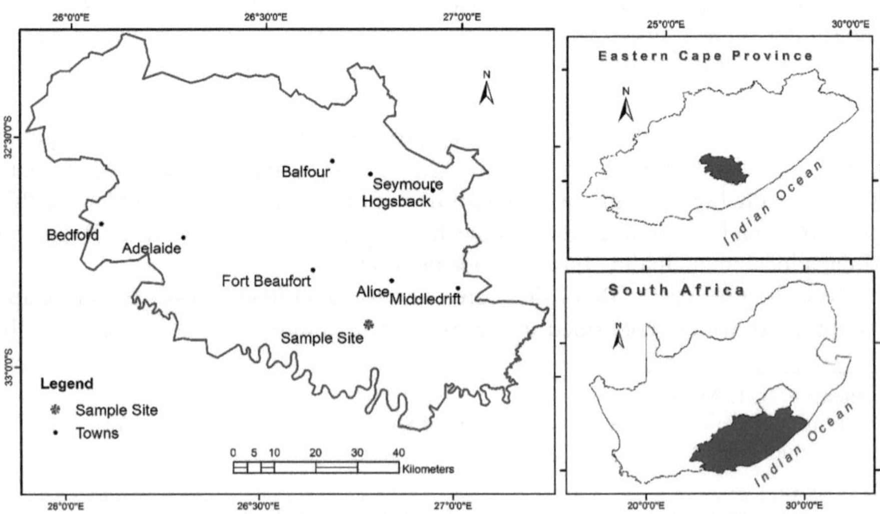

Fig. 6.1 Map of small towns in the Raymond Mhlaba Local Municipality, 2022. (Source: Raymond Mhlaba Local Municipality, 2022)

growing backlog; second, poorly built and located low-income housing units, and third, low-income housing provision that fails to promote spatial integration, primarily through infill development.

6.2 Small Towns and Housing Provision: A South African Historical Overview

The existence of small towns is not unique to South Africa but is a global phenomenon. From a South African perspective, there is a historical context to the role of small towns, including the general state of infrastructure, particularly housing in these towns. As argued by Redding (1992), the significance and role of small towns in South Africa dates far back into the period prior to the 1950s. With large cities, such as Durban, Johannesburg and Cape Town not being typical of South Africa prior to the economic growth and development of the 1950s and 1960s (Manona, 1988), it was only in small towns where (through a strictly controlled and limited interaction with their colonisers), black South Africans in particular could be provided with their only urban experience (Redding, 1992). The historical significance of small towns is confirmed by a significant number of citizens across developing countries (South Africa included), who continue to live in these towns and their surrounding rural hinterland (Gibb & Nel, 2007). The significance and global standing of small towns is also confirmed by Satterthwaite and Tacoli (2006) who argued that more than half the population of Africa, Asia and Latin America resides in small towns with a total population of fewer than 50, 000 people. Hence, it should not come as a surprise that there is a significant number (about 500) small towns in South Africa (Van Niekerk & Marais, 2008; Centre for Development and Enterprise, 1996). Like elsewhere in the world, particularly in developing countries, the functions of small towns in South Africa are closely related to and to a certain extent, reflective of rural-urban linkages and interactions. Other than being service centres for the surrounding rural hinterland, one other important role of small towns is their historical role of attracting rural migrants who would otherwise migrate to big cities and towns with more economic and employment opportunities (Agergaard et al., 2018).

Despite playing such an important historical role(s), small towns in post-apartheid South Africa continue to endure not only neglect by researchers and academics (Nel, 2005; Gibb & Nel, 2007), but political and economic neglect by the government as well (Van Niekerk & Marais, 2008; Agergaard et al., 2018). Subsequent to neglect by researchers and academics, communities in small towns have rarely become the focus area of research on housing and neighborhood issues (Skobba et al., 2020). With regard to neglect by government, such is confirmed by the South African government's failure to formulate a differentiated/targeted national policy on small town development (Van Niekerk & Marais, 2008). Hence, as a result of such neglect and subsequent policy failure, South African small towns are in a state of economic decline in general (Gibb & Nel, 2007). The general economic decline and neglect of small towns by national government in particular, seems to have manifested itself in many ways. First, is through poor governance and inadequate infrastructural development across various sectors, public or low-income housing being one such sector. Confirming this is the argument by Merten (2003) on how, subsequent to their struggle to provide socio-economic development, many small towns are found to be unable to successfully address, in particular, housing shortage and the related service delivery issues. Consequently, most if not all former black township areas found on the periphery of every big city and small town in South Africa are known for amongst other things, the proliferation of informal structures and backyard dwellings (Shackleton & Blair, 2013). There is an estimated national figure of 2600 informal settlements in the country (Maqhina, 2021), while the national housing backlog amongst poor households qualifying for state-funded, low-income housing increased from a 1.4 million housing backlog in 1994 (Department of Housing, 1994) to the current figure of just over 2.2 million housing units in the country (Chakwizira, 2019). This, despite the government's housing subsidy programme (commonly known as RDP houses in South Africa), delivering just over 3 million subsidised housing units since the advent of democracy in 1994 (Chakwizira, 2019). It may therefore be appropriate to argue that inadequate provision of low-income public housing, particularly for black urban communities in small towns, was not a long-standing development challenge for the apartheid government only (see Redding, 1992), but continues to remain an inherent challenge facing a democratic state, post-1994 (Nel, 2005). Existing evidence shows how the current housing landscape and related issues in most small towns or rural local municipalities governing small towns continues to be characterised by a growing housing backlog, poorly built low-income houses and incomplete low-income housing projects, amongst other things. For instance, subsequent to a shortage of state-funded RDP housing in the small town of Makhado (originally named Louis Trichardt) in Limpopo province, there has been a proliferation of informal settlements through illegal land occupation on the periphery of this town (Nengovhela & Van Zyl, 2021). Similarly, illegal land occupation continues to contribute not only to the housing backlog but to a growing number of informal settlements in the small town of Ventersdorp (North West province) as well (Motse, 2017). Other than informal settlements, contributing to a growing housing backlog and inadequate shelter in small towns, is poor workmanship on RDP houses by the state's appointed building contractors. For instance, in the small town of Ficksburg (Free State province), the government had to terminate a contract and stop a house-building project for some 70 houses due to poor quality work by the contractor (Sapa, 2013). Another small town where the urban housing landscape is characterised by poor workmanship and thus, a housing backlog, is Stillwater near Riverton in the Northern Cape province. Beneficiaries of an RDP housing project which was completed in 2014 are now blaming (amongst other things), collapsing housing units or cracking walls of their houses on poor workmanship by the state's appointed contractor (Hlakotsa, 2022). Growing housing backlog due to poor workmanship by the state's appointed contractors was also experienced by residents of Namakgale township, in the small mining town of Phalaborwa (Limpopo province). Evidence shows that poor workmanship led to several RDP housing units collapsing in the Tshelang-Gape section in Namakgale township; an incident which left the Limpopo government with no option but to issue a noncompliance certificate to the contractor (Matlala, 2020).

In most small towns, the lack of adequate housing is not a challenge facing poor households only, but middle-class people as well. As argued in the literature, there is an unmet high demand for middle-class housing in most small towns, especially those with some potential for economic growth and development (Kitchen, 2008). This is as a result of a lack of capacity by municipalities in these small towns to either afford investment in low-cost rental housing and/ or social housing stock or attracting the private sector to invest in the housing opportunities (Ntema, 2021; Kitchen, 2008). In summary, it may be appropriate to argue that the untransformed housing landscape in most small towns can largely be attributed to an inability by these towns and their respective local municipalities to overcome the legacy of apartheid spatial planning or spatial injustices (Van Niekerk & Marais, 2008). For instance, the perpetuation of urban sprawl through the peripheral location of state-funded, low-income public housing development is a common phenomenon, not only in big cities but in most small towns as well. This in the main, could be attributed to a fourfold institutional challenge. First, is the lack of institutional capacity in most municipalities governing small towns to put in place effective land use management plans (Kitchen, 2008). Second, is the failure by some municipalities governing small towns to implement their policy on integrated human settlement development (Ntema, 2021). Third, is the failure by government to formulate and promulgate a targeted/differentiated national policy on small towns (Van Niekerk & Marais, 2008). Fourth, is the failure by the National Spatial Development Perspective (NSDP) to articulate a clear stance on small towns (Oranje et al., 2008). This despite its (NSDP) objective being to assist government in answering (through a concrete plan) a critical question on where should government direct its investment and development initiatives to ensure inclusive, sustainable and maximum economic development and growth or impact. Hence, Van Niekerk and Marais (2008), are of the view that only a few small-town, development success stories exist in South Africa. The housing backlog and thus, untransformed spatial planning in most small towns discussed above, should be understood within a national context. Despite completion (at least on paper), of just over 3 million state-funded, low-income housing units in the country since 1994 (Chakwizira, 2019), about 1.9 million of these housing units remain incomplete due to amongst other things, incomplete or abandoned housing projects (Maqhina, 2021). To complete these housing units that are left by incompetent contractors at the foundation phase or wall plate, the national government would have to raise no less than R300 billion. With 1.9 million incomplete housing units, the question should be whether the current official figure of just over a 2.2 million housing backlog in the country is a true reflection of the real situation on the ground.

6.3 Small Towns and Housing Provision in Eastern Cape Province

The discussion in this section will focus exclusively on the provision of state-funded, low-income (RDP) housing and the experiences of communities in poor, rural local municipalities and small towns under their jurisdiction. As indicated in the introduction above, the Eastern Cape province is one of the most rural and poorest provinces in South Africa. The province comprises 39 local municipalities with only two of these being metropolitan areas, Nelson Mandela Bay Metro and Buffalo City Metro, while the rest are predominantly small, poor, rural local municipalities (Auditor-General, 2021). Like elsewhere in the country, most municipalities (local and district levels) in the Eastern Cape province continue to grapple with undoing the socio-economic and spatial disparities mostly inherited from the apartheid government in 1994. While past apartheid spatial planning and housing policies contributed to the current context, it is however, appropriate to argue that continued socio-economic and spatial disparities in most municipalities across the province should, in the main, be understood within the broader context of a general lack of implementation of various policies and accountability by government post-1994 era. As noted in various annual reports by the Auditor-General, the culture of irregular expenditure by municipalities in the province seems commonplace and institutionalised. For some time now, there has been a steady increase in irregular expenditure by municipalities in the province; the latest evidence shows an increase from 2.5 billion for the financial year 2018/19 to 3.1 billion for the 2020/21 financial year (Auditor-General, 2021, 2022). The main contributor to these financial irregularities is amongst other things, a widespread failure by most municipalities, including the small, rural Raymond Mhlaba local municipality to put in place proper systems to identify and disclose all irregular expenditure as required by section 125(2)(d) of the Municipal Finance Management Act (MFMA) (Raymond Mhlaba Local Municipality, 2021; Auditor General, 2021). The poor financial management and lack of accountability in municipalities across the province have negatively affected the provision of basic services including housing provision, particularly in small towns governed by poor, rural local municipalities.

From a housing provision perspective, the housing landscape in the province is dominated by formal housing. There are currently 72.4% of households residing in formal housing across the province (Stats SA, 2021). However, given the rural nature of the province and the high level of poverty, particularly in the small, rural towns that currently dominate the provincial urban landscape (see also Westaway, 2012), it does not come as a surprise to see 21.6% of households residing in traditional housing, a figure way above the national average of 4.2%. Notwithstanding the high percentage (72.4%) of households residing in formal housing, there is currently about 5.4% of households residing in informal housing across the province (Stats SA, 2021). In summary, there is a combined figure of just over a quarter (26.4%) of households in the province that reside in dwellings other than formal dwellings (Stats SA, 2021). This translates into a 400,000 backlog of housing units in the entire province (Hlakotsa, 2022). Worth noting is that most of these households facing the housing backlog reside in small, rural towns and local municipalities outside the two metropolitan areas. Confirming this view, the literature shows how, despite their long-standing and common history of being depressed and collapsing economically (see also Raasch, 2001), small, rural towns (particularly in the Eastern Cape province) have become known for their long-standing housing backlogs and thus, a strong demand for state-funded, low-income housing development (Bank et al., 2010).

While the low number (8.5%) of households residing in rented dwellings in the province could possibly be indicative of the high rate of home ownership, it could also be attributed to a shortage of rental housing stock in the province (see discussion below). There is a steady decline in the number of households with access to piped or tap water in their dwellings, off-site or on-site since 2014; the figures show a decline from 78.7% in 2014 to 71.0% in 2021, while the current 43.9% of households with pit latrine toilets is the highest in the country (Stats SA, 2021). The possible implication of this could be that other than households in informal settlements, there could still be state-funded, low-income housing units built without access to these basic services. By all accounts, this goes against the notion that "adequate housing is the sum of a number of considerations including location, basic infrastructure, affordability, sustainability and right to tenure amongst others" (Pottie, 2004:610). Consequently, 32.6% and 34.1% of households in 2016 rated their RDP/state-funded dwellings as 'poor' and 'average' respectively, while only 33.3% rated them as 'good' in this province (Stats SA, 2016). Similar to no formal housing, most households who rated their state- funded housing as 'poor' reside in small towns, usually found in poor, rural local municipalities outside metropolitan areas, with the highest number being in the poor, rural district municipality of Amathole (Stats SA, 2016). This is confirmed by 49.5% of households in Amathole district rating their RDP/state-funded dwellings as 'poor', while a mere 19.7% rated them as 'good' in 2016 (Stats SA, 2016). This rating of 49.5% is not only higher than the provincial average of 32.6% but of the rest of the other five district municipalities, including two metropolitan areas.

The study by Manomano and Tanga (2018) further confirms widespread dissatisfaction amongst beneficiaries of the state-funded, low-income housing projects in small towns that are governed by the four, rural local municipalities of Mbashe, Mnguma and the former Nkonkobe and Nxuma local municipalities. Beneficiaries of state-funded RDP housing in various small towns across these four local municipalities expressed an average rate of 90% dissatisfaction about housing structural defects, ranging from leaking roofs, poorly built walls, poorly fitted windows and doors, to the poor quality of building materials in general (see also Manomano, 2013). Widespread dissatisfaction about housing structural defects amongst beneficiaries of housing projects in small towns could further be confirmed by a qualitative study conducted by Ngcayi (2013) in the small town of Peddie, which is governed by the small, rural Nqgushwa local municipality. In this study, evidence from in-depth interviews and focus group discussions show high levels of dissatisfaction about the growing housing backlog, and a lack of transparency by both municipal officials and ward councillors, when housing waiting lists are compiled. For example, there are problems regarding the peripheral location of RDP housing, poor workmanship by incompetent contractors appointed by government, and RDP housing with structural defects, such as small size, cracks and leaking roofs. Both local government officials and community members in this small town of Peddie blame the growing housing backlog on inadequate funding by the provincial and national government spheres and a lack of internal capacity at local level. In 2008 a similar study focusing on housing provision was conducted by Kruuse (2008) in the six small towns of Alicedale and Riebeeck East (small rural Makana local municipality); Peddie and Hamburg (small rural Nqgushwa local municipality), and Cala and Elliot (small rural Sakhisizwe local municipality). According to this study (Kruuse, 2008), all of these six small towns and their respective local municipalities are faced with a variety of common threads, related particularly to housing provision. Some of the common housing related challenges that were found to be facing these six small towns in 2008 include the housing backlog, complaints about state-funded RDP housing projects that have over a number of years remain blocked or incomplete, and those which lie dormant (Kruuse, 2008). The housing backlogs and the general poor state of housing in these six small towns and their local municipalities in general, could in the main, be attributed to the following: absence of housing strategies or schemes and housing sector plans; a lack of internal capacity at local level; a lack of support and poor communication between municipal officials responsible for housing provision and their provincial counterparts, and the non-implementation of key strategic planning documents, such as IDPs. For instance, despite being the only strategic planning documents that outline and make some reference to matters relating to housing provision, in 2008 the IDPs in Nqgushwa and Sakhisizwe local municipalities were criticised for being *'nothing more than hollow statements of good intentions'* and for being *'councillors' wish lists'* which were never implemented in response to the growing housing shortage in these poor communities (Kruuse, 2008). There is also no proper alignment between IDPs in these two local municipalities and the province's plans in respect of housing provision (Kruuse, 2008). Administratively, evidence from this study further shows that internal housing units or directorates remain inefficient and ineffective due to a lack of external support by the provincial department of housing and a lack of internal capacity. For instance, with the exception of the Makana local municipality, in both Nqgushwa and Sakhisizwe local municipalities, there is only one under-resourced housing officer in each municipality. Other than housing, the single housing officer in the Sakhisizwe local municipality was also responsible for matters related to Local Economic Development (LED), Integrated Development Process (IDP), and Land and Sports and Recreation, while one in the Nqgushwa local municipality was also responsible for Land and Town Planning and IDP (Kruuse, 2008). The current challenges related to housing provision in various poor, rural local municipalities and their small towns across the province, are not without a historical context. Historically, the literature shows overall, how the provision of housing in the Eastern Cape province has had little or no regard for integrated development planning and has been seen as a stand-alone deliverable (Bank et al., 2010). Next is the discussion on how the poor, rural Raymond Mhlaba local municipality (which is located in rural Amathole district municipality) performed in transforming the socio-spatial make up in its seven small towns through infrastructural development, particularly housing.

6.4 Provision and State of Housing in the Raymond Mhlaba Local Municipality

The Raymond Mhlaba local municipality is predominantly rural, with 72% of its total population residing in rural villages (Raymond Mhlaba Local Municipality, 2022). The municipality consists of 23 wards across all the seven small towns of Alice, Fort Beaufort, Bedford, Adelaide, Middledrift, Hogsback, and Seymour and their rural hinterland, as shown in Fig. 8.1 below (Raymond Mhlaba Local Municipality, 2022). While all these seven small towns are service centers to the surrounding rural communities (Raymond Mhlaba Local Municipality, 2018a), the following exceptions are worth noting- Alice further retains its historical role of being a cultural and educational centre (Mamba & Isabirye, 2015) and is also a political seat of Raymond Mhlaba local municipality (Raymond Mhlaba Local Municipality, 2018a), while Fort Beaufort and Hogsback further serve as the administrative capital of the Raymond Mhlaba local municipality and sub-regional center respectively (Raymond Mhlaba Local Municipality, 2018a).The total population in this local municipality is around 163,000 people (Raymond Mhlaba Local Municipality, 2022). However, the municipality has experienced a population decline since 1996. Contrary to the current population of around 163,000 people, the municipality had 168,953 people in 1996 (Raymond Mhlaba Local Municipality, 2022; Stats SA, 2013). Such a decline should probably be understood within the context of what Hoogendoorn & Visser (2016:95) termed "a phase of rural depopulation and small town decline" that countries, particularly those in the global south usually undergo.

Notwithstanding both the provision of state-funded, low-income housing opportunities in the local municipality to just over 10,000 households since 1994 and a high percentage (66.1%) of households currently residing in formal housing, the housing backlog and poor state of existing housing stock remain a challenge in the Raymond Mhlaba local municipality (Raymond Mhlaba Local Municipality, 2018a). Evidence in all the seven small towns shows a growing demand for state-funded, low-income housing stock for the purpose of home ownership and rental, in the form of social housing (Raymond Mhlaba Local Municipality, 2022). On one hand, evidence shows that across the seven small towns there is a combined figure of about a 6700 housing backlog, with about 3310 households residing in informal settlements, while close to 1000 households reside in informal structures in backyards across the local municipality (Raymond Mhlaba Local Municipality, 2022). Noteworthy, is a sudden increase in the number of households residing in informal settlements from 861 households in 2018 to the current 3310 households (Raymond Mhlaba Local Municipality, 2018a, 2022). This sudden increase should, amongst other things, be understood within the context of covid-19 which (between 2020 and 2021) led to nationwide, sporadic land grabs. For example, the city of Cape Town was forced to deal with a staggering 260 incidents of alleged illegal land occupation between April and July 2020 lockdown (Evans, 2020). As argued by Reinders (2020), the sudden development of informal settlements with names such as 'Covid', 'Sanitizer' and '19' (as in covid-19) in and around big cities like Cape Town during the lockdown is a confirmation of a staggering rise in "land invasions" since covid-19 hit the country. Confirming the growing housing demand due to backlogs in Raymond Mhlaba local municipality could, on one hand, be the demands made particularly by residents in previously poor, black communities, such as those residing in wards 2, 10, 11, 18, who demanded the construction of RDP houses in the Maplangeni area, in the Gomoro and Mathole areas, in the Dyamala area, and in the Njani and Sheshegu areas respectively, during community based planning (Raymond Mhlaba Local Municipality, 2018b). On the other, evidence for the inadequate and poor state of existing housing shows a significant number of about 9400 households residing in traditional dwellings, which include huts made of traditional material across all the small towns in the Raymond Mhlaba local municipality (Raymond Mhlaba Local Municipality, 2022). The growing housing backlog in the Raymond Mhlaba local municipality is, to a large extent, indicative of a greater challenge facing the entire province, which has about a 400,000-housing backlog (Hlakotsa, 2022).

As shown above, the housing backlog manifests itself through amongst other things, a growing number of informal settlements and backyard dwellings, with the highest concentration being in Fort Beaufort, followed by Adelaide, Bedford, Middledrift, Seymour and Hogsback. Alice town seems exceptional, with the lowest number of informal settlements; this, despite the growing housing demand and shortage of accommodation for students and academic staff in both the University of Fort Hare and Lovedale FET College, including professionals employed in public institutions, such as local clinics and the Victoria Hospital (Raymond Mhlaba Local Municipality, 2022; 2018a; 2018b; Ntema, 2021). Thus, it does not come as a surprise to see an estimated figure of about 3000 of the current 6700 housing backlog found in Alice town (Raymond Mhlaba Local Municipality, 2018a, 2022). The current housing landscape in the municipality is further marred by a significant number of housing units that require rectification in various housing projects across all the towns. Despite the recent completion of the rectification process in Seymour and Middledrift, the local municipality is yet to fund and undertake rectification in about 638 housing units in Fort Beaufort Hillside, 300 housing units in Kanana, and about 662 housing units in the Newtown area (Raymond Mhlaba Local Municipality, 2022). This is to be expected given the similar challenge of the high number of RDP housing that require renovation in the entire province. In the 2021 financial year, the province had to set aside R100 million for restructuring or rebuilding poorly constructed RDP housing (Hlakotsa, 2022). If issues related to housing backlog, the poor state of existing housing stock, incomplete projects and rectifications are anything to go by, it demonstrates that housing is amongst the major community grievances raised by communities in all seven small towns during community-based planning by the municipality (Raymond Mhlaba Local Municipality, 2018b). As shown in the discussion above, these and other housing related issues are not unique to the Raymond Mhlaba local municipality but to the entire district municipality of Amathole and the Eastern Cape province by extension.

6.4.1 Housing Provision and Socio-Spatial Transformation in the Raymond Mhlaba Local Municipality

It is our view that similar to the apartheid government which used housing development to promote its policies on separate development, racial segregation, influx control and orderly urbanisation (Harrison & Todes, 2015; Mabin, 1991; Platzky & Walker, 1985), the democratic government including the Raymond Mhlaba local government, could equally use state-funded housing development as a tool to promote its policies built generally on the principle of inclusive development. However, despite featuring prominently in both national policies – the 1994 White Paper on Housing; the 2004 Breaking New Grounds (BNG); the municipality's strategic planning documents, such as the Spatial Development Framework (SDF) and the Integrated Development Plan (IDP), it would seems that the provision of integrated state-funded, low-income housing development in the Raymond Mhlaba local municipality remains largely lip service (Ntema, 2021; Raymond Mhlaba Local Municipality, 2018a). Notwithstanding strides made in housing provision since 1994, the housing landscape in the seven small towns that constitute this municipality, continue to mirror what Thakrar (2018) termed 'spatial disconnect'. This view could, to a large extent, be confirmed by the dominant application of two words seen as a possible solution: *'infill housing'* and *'integrated human settlement'* in the Raymond Mhlaba SDF situational analysis document. The word *'infill housing'* appears no fewer than 30 times, while *'integrated human settlement'* appears no fewer than 40 times in this strategic planning document (Raymond Mhlaba Local Municipality, 2018b). Through the internal diagnosis presented in the 2022 IDP document, it would seem that from one small town to another, the Raymond Mhlaba local municipality continues to grapple with unintended spatial consequences of either existing state-funded, low-income housing projects or existing housing backlog. A number of historical socio-spatial disparities that could have been addressed through housing provision and development, are worth noting.

First, through a SWOT analysis done by the municipality in the 2018/2019 financial year, it was found that the current spatial disparities across small towns in the Raymond Mhlaba local municipality cannot be blamed for the shortage of state-owned land, particularly in key strategic locations. Despite communal land ownership in almost all the rural villages (Raymond Mhlaba Local Municipality, 2018b), evidence shows that a bulk of land in the seven towns across Raymond Mhlaba local municipality is owned by the state- for example, more than 90% of land in Alice is owned by the state (Ntema, 2021). Instead, the internal diagnosis revealed how, for the longest time, the local municipality has paid lip service to the formulation and implementation of the integrated and sustainable human settlement development plan (Raymond Mhlaba Local Municipality, 2018b; Ntema, 2021). Consequently, a failure to formulate and implement the integrated and sustainable human settlement development plan has manifested itself in various spatial disparities across towns in the local municipality. Amongst these spatial disparities is the widespread lack of town-based spatial in-fill development or densification; the lack of in-situ upgrading projects, and continued urban sprawl coupled with low-density development (Raymond Mhlaba Local Municipality, 2018b, 2022). If (as shown in Fig. 6.2) the current situation found in both a primary development corridor along the R63 (Bedford, Adelaide, Fort Beaufort, Alice, and Middledrift), a secondary development corridor along the R67 (Seymour and Fort Beaufort), and a tertiary development corridor along the R345-R344 (Alice, Hogsback and Adelaide), meant to ultimately revitalise all these seven small rural towns is anything to go by, it may be appropriate to argue that spatial disconnect exists between towns as well. This is confirmed by a concern raised in the IDP about the spatial disconnection between Middledrift and key strategic towns, such as Alice and King William's town. At the centre of this spatial disconnect between towns is the absence of either state-funded or privately funded housing development along the course of strategic provincial roads or corridors of development, such as the R63 between Alice and Fort Beaufort which is 20 km west and Middledrift which is 20 km east. This, despite such housing development being an integral part of the municipality's plan on all the

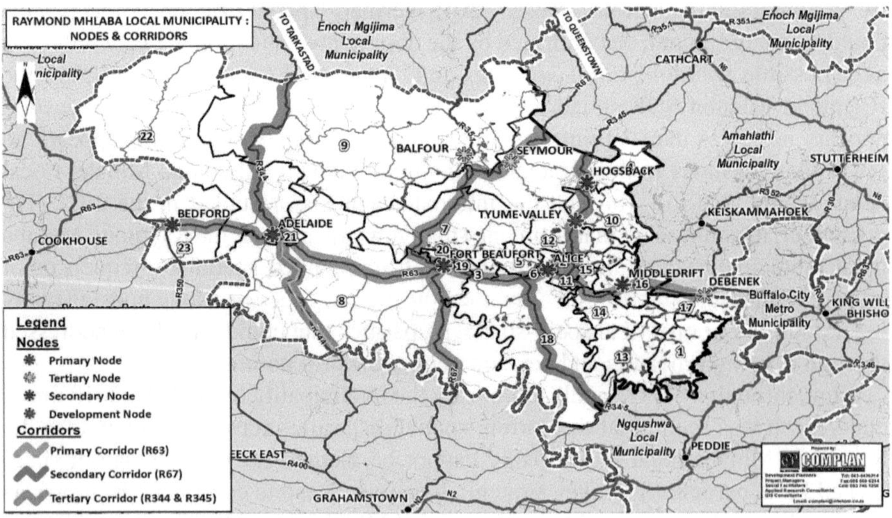

Fig. 6.2 Corridors of development and spatial integration in Raymond Mhlaba Local Municipality, 2021. (Source: Raymond Mhlaba Local Municipality, 2021)

corridors included in the IDP and SDF documents (Raymond Mhlaba Local Municipality, 2018b, 2022).

Further perpetuating the historical spatial inequalities and the failure to go beyond mere acknowledgement of housing provision as a possible solution, could be the absence of any concrete small-town revitalisation strategy or programme in the local municipality. Evidence shows that other than a small-town revitalisation strategy or programme for Alice, although debatable (see also Ntema, 2021), none of the other six towns has an existing revitalisation strategy or programme (Raymond Mhlaba Local Municipality, 2022). This, despite socio-economic and spatial challenges in all these seven towns being similar.

Several internal and external factors seem to have (directly or indirectly) contributed to the institutional failure by the local municipality to formulate and implement an integrated and sustainable human settlement development plan. First, is the failure by the local municipality to build internal administrative capacity required to formulate or put in place a housing sector plan, as well as a directed and compact growth strategy (Raymond Mhlaba Local Municipality, 2018b, 2022). With regard to issues related to 'compact growth strategy', there is acknowledgement by the local municipality that Alice town requires a high density student hotel development in close proximity to the University of Fort Hare and within walking distance of the CBD, together with a high density housing development in the corridor (R63 road) between this town and Fort Beaufort. Both Fort Beaufort and Adelaide towns require high density, mixed-use development towards the Grahamstown-Adelaide junction and along the R63 road in and out of Bedford town respectively. From an integrated and sustainable human settlement development perspective, a lack of 'compact growth strategy' could be confirmed by a call in the IDP and SDF for in-fill housing or residential development on state owned vacant, derelict, underdeveloped and underutilised buildings and land parcels in strategic locations, such as the CBDs in Alice, Fort Beaufort, Hogsback, Adelaide and Bedford towns (Raymond Mhlaba Local Municipality, 2018b, 2022). For instance, in Alice town there is clear evidence for the utilisation of part of the eGolfini area, Alice west and CBD to counter both urban sprawl and segregated residential areas, while promoting in-fill housing and residential development. Second, is the failure to put in place effective policy and programmes to address land ownership and tenure issues, particularly in rural communities adjacent to more rural towns, such as Middledrift, Seymour and Bedford (Raymond Mhlaba Local Municipality, 2018b, 2022). Evidence shows how land disputes launched by traditional leaders, particularly in villages adjacent to Middledrift, Seymour and Bedford towns remain a hindrance to the government programme on rural housing provision, especially in so-called 'prioritised villages'. Hence, a resolution by local municipalities should be implemented to ensure that traditional leaders are mobilised into supporting settlement planning initiatives (Raymond Mhlaba Local Municipality, 2018b). Third, is the growing number of blocked housing projects and those requiring rectification since 2018. Across all seven towns, the Raymond Mhlaba local municipality recorded a figure of 3429 housing units that required rectification in the 2018/19 financial year. A further 2422 housing units formed part of the blocked housing projects across the seven towns in the same financial year. For instance, towns hit hard with the blocked housing projects include, Kuntselamanzi and Hillcrest in Alice with 818 units, Hillsides in Fort Beaufort with 638 units, followed by Seymour with 461 units (Raymond Mhlaba Local Municipality, 2018b, 2022). Fourth, is the administrative failure by the municipality to conduct a comprehensive land audit, the absence of a wall-to-wall scheme, and the absence of a land use scheme (Raymond Mhlaba Local Municipality, 2022; Ntema, 2021). For instance, as part of the CBD revitalisation, the local municipality has acknowledged the urgent need for urban design and a land use planning/scheme, in order to create a spatial connection between the CBD in Adelaide town and a spatially dislocated township area. Similar concerns are being raised with regard to spatial disconnect between other small towns and their respective historical black township areas: Alice and Fort Beaufort towns, and eNtselamanzi and eBhofolo township areas respectively being one such example (Raymond Mhlaba Local Municipality, 2018b, 2022). Seymour town is no exception in this regard as there is the acknowledgement that this town requires proper spatial planning and land use management/planning, in order to promote sustainable social, economic and environmental development (Raymond Mhlaba Local Municipality, 2022). Fifth, is the failure by the local municipality to define the 'urban edge or fringe' in almost all of the seven towns. Consequently, the provision of housing in these seven small towns continues to perpetuate inherent, apartheid spatial inequalities through urban sprawl, coupled with low-density development. Some examples in this regard include a need to delineate the settlement edge in Adelaide town, and the Seymour and Bedford towns. The perpetuation of urban sprawl through state-funded, low-income housing development in these small towns does not only undermine government's national programme on spatial redress but goes against post-1994 policy discourse as well. For instance, in terms of the 1994 White Paper on Housing and 2004 BNG, any state-funded, low-income housing development should assist in "promoting integrated communities situated in areas allowing *progressive access* to portable water; adequate sanitary facilities; waste disposal; economic opportunities; educational and social amenities" (Department of Housing, 1994:19) and "promoting integrated and sustainable human settlement development" (Department of Housing, 2004:25). Sixth, is a policy gap or misalignment between the localised Raymond Mhlaba SDF and the Amathole district SDF and the Eastern Cape Provincial SDF. This misalignment could amongst others be attributed possibly to poor communication due to weak intergovernmental relations between these three spheres of governance (see also Ntema, 2021). Seventh, is the failure by the local municipality to forge any formal partnership with the private sector and institutions of higher learning (University of Fort Hare and Lovedale FET College) for a possible joint investment in housing development projects that could respond to the growing accommodation demand, particularly for middle-class people in towns, such as Fort Beaufort, Alice, Bedford and Adelaide (Raymond Mhlaba Local Municipality, 2018b, 2022). Such investment in housing development could also include off-campus private accommodation for students and staff members at the University of Fort Hare and Lovedale FET College (Ntema, 2021). A review of the recent IDP document does not make mention of any existing formal partnership or joint venture between the Raymond Mhlaba local municipality and the private sector that in the short to long term, could assist in responding to growing housing demands for home ownership and rental stock, mainly amongst middle-class groups (Raymond Mhlaba Local Municipality, 2022; Ntema, 2021).

6.5 Concluding Remarks

The discussion above, makes it appropriate for this chapter to conclude that the socio-economic and spatial challenges currently characterising the housing landscape in most post-apartheid small towns are inherently historical in nature. It seems that a tendency to disregard the principle of 'integrated development planning' as an essential part of any housing provision, is not a practice unique to the Raymond Mhlaba local municipality, but to most local municipalities in the Eastern Cape province. Despite the strides made, the provision of state- funded, low-income (RDP) housing in small towns that constitute the Raymond Mhlaba local municipality, could in the main, be criticised for being inadequate and less effective in promoting spatial redress within and between towns. Therefore, one could argue that contrary to an emphasis on policy shift from housing development to sustainable and integrated human settlement development by the 2004 BNG national policy and municipal strategic planning documents, such as the IDP and SDF, residents in the seven small towns across the Raymond Mhlaba local municipality seem to have experienced (although debatable) access to housing, instead of access to integrated and sustainable human settlement. The current shortage of housing and the associated continuation of spatial disparities in the Raymond Mhlaba local municipality should be understood within a threefold scenario. First, is the general lack of political will and capacity at local level to use both state-owned derelict buildings and vacant land in all strategic locations to reduce growing housing backlog while promote *spatial infill.* Interestingly, the word *'spatial infill'* has not only been mentioned under each of the seven towns but has also been mentioned no fewer than 30 times in the Raymond Mhlaba SDF situational analysis document. Second, is the high unemployment rate which is around 50% in the local municipality. Third, is the high poverty levels, with just over 70% of the population declared as poverty stricken (Raymond Mhlaba Local Municipality, 2022). The growing housing backlog, chiefly amongst the middle-class in Alice, Fort Beaufort and Adelaide, and the student population concentrated mainly in Alice, and Fort Beaufort to some extent, is likely to increase in the future, unless the Raymond Mhlaba local municipality formally adopts and includes in their IDP, a concrete plan on how to attract active role-players and investment by the private sector (public-private partnership) in rental housing stock. Such a concrete plan should also outline how the local municipality intent to create enabling environment for a joint-investment by government and the two institutions of higher learning-University of Fort Hare and Lovedale FET College in off-campus private accommodation for students and staff members.

References

Agergaard, J., Tacoli, C., Steel, G., & Ortenblad, S. B. (2018). Revisiting rural-urban transformations and small town development in sub-Saharan Africa. *European Journal of Development Research, 31*, 2–11.

Auditor-General. (2021). *Consolidated general report on the local government audit outcomes.* Office of Auditor-General.

Auditor-General. (2022). *Consolidated general report on the local government audit outcomes.* Office of Auditor-General.

Ballard, R., & Steyn, M. (2013). Diversity and small town spaces: Twenty years into post-apartheid South African democracy. *Diversities, 15*(2), 1–66.

Bank, L., Makubalo, L., & Maqasho, L. (2010). *The Eastern Cape basic services delivery and socio-economic trends series: Housing delivery and state subsidies.* Fort Hare Institute of Social and Economic Research.

Centre for Development and Enterprise. (1996). *South Africa's small towns: New strategies for growth and development.* Centre for Development and Enterprise.

Chakwizira, J. (2019). Low-income housing backlogs and deficits "Blues" in South Africa. What solutions can a lean construction approach proffer? *Journal of Settlements and Spatial Planning, 10*(2), 71–88.

Department of Housing. (1994). *White paper on housing: A new housing policy and strategy for South Africa* (Government Gazette No. 16178). Department of Housing.

Department of Housing. (2004). *"Breaking new ground" A comprehensive plan for the development of sustainable human settlements.* Department of Housing.

Evans, J. (2020, August 3). *260 lockdown 'land invasions' and counting for city of Cape Town.* News24.

Gibb, M., & Nel, E. (2007). Small town redevelopment: The benefits and costs of local economic development in Alicedale. *Urban Forum, 18*, 69–84.

Harrison, P., & Todes, A. (2015). Spatial transformations in a 'loosening state': South Africa in a comparative perspective. *Geoforum, 61*, 148–162.

Hlakotsa, N. (2022, August 10). *Stillwater residents claim RDP houses built in 2014 are falling apart.* SABC News.

Hoogendoorn, G., & Visser, G. (2016). South Africa's small towns: A review on recent research. *Local Economy, 31*(1–2), 95–108.

Kitchen, F. (2008). *Land use management, spatial planning and the land market in small towns.* Academy of Science of South Africa.

Kruuse, H. J. (2008). *Life in the suburbs after Grootboom: The role of local government in realising housing rights in the Eastern Cape* [Master's theses]. Rhodes University.

Mabin, A. (1991). The dynamics of urbanization since 1960. In M. Swilling, R. Humphries, & K. Shubane (Eds.), *Apartheid city in transition* (pp. 33–47). Oxford University Press.

Mamba, M. S. N., & Isabirye, N. M. N. (2015). A framework to guide development through ICTs in rural areas in South Africa. *Information Technology for Development, 21*(1), 135–150.

Manomano, T. (2013). *The perceptions of the Reconstruction and Development Programme (RDP) housing beneficiaries in South Africa on the extent to which the project meet their housing needs: The case of Golf Course Estate in Alice town, Eastern Cape Province* [Master's theses]. University of Fort Hare.

Manomano, T., & Tanga, T. (2018). Housing needs: The quality and quantity of housing provided by the government for the poor in the Eastern Cape province in South Africa. *Social Work, 54*(1), 19–36.

Manona, C. (1988). Small town urbanization in South Africa: A case study. *The African Studies Review, 31*(3), 95–110.

Maqhina, M. (2021, November 27). *Incomplete RDP houses stand at almost two million and fixing them could cost R300bn.* Independent Online (IOL).

Matlala, A. J. (2020, September 2020). Limpopo RDP houses 'fall apart' after storm. *The Citizen.*

Merten, M. (2003, April 18). *Small towns struggle with delivery*. Mail & Guardian.
Motse, O. (2017, August 15). *North West's poor delivery of RDP houses criticised.*.
Nel, E. L. (2005). Local economic development in South African small towns. In E. L. Nel & C. M. Rogerson (Eds.), *Local economic development in the developing world: The experience of Southern Africa* (pp. 253–266). Transaction Press.
Nel, E. L., & Hill, T. R. (1998). Demographic trends in the Eastern Cape Province of South Africa: Development and planning implications. *Urban Forum, 9*(1), 130–141.
Nengovhela, K., & Van Zyl, A. (2021). You promised us RDP housing, not an auction, say shack protesters. *Limpopo Mirror* (p. 2).
Ngcayi, Q. Q. (2013). *Service delivery, the state and NGOs – The case of Peddie extension, Eastern Cape* [Master's theses]. Nelson Mandela Metropolitan University.
Ntema, J. (2021). Alice: Socio-spatial change in a small university town. In A. Lemon, R. Donaldson, & G. Visser (Eds.), *South African urban change three decades after apartheid, homes still apart?* (pp. 233–247). Springer Nature.
Ntema, J., & Venter, A. (2016). Polokwane. In L. Marais, E. Nel, & R. Donaldson (Eds.), *Secondary cities and development* (pp. 125–139). Routledge.
Oranje, M., Van Huyssteen, E., & Meiklejohn, C. (2008). *The National Spatial Development Perspective (NSDP) and assumptions on small town economic investment by government.* Academy of Science of South Africa.
Pauw, K. (2005). *The profile of the Eastern Cape province: Demographics, poverty, inequality and unemployment*. Provide Project.
Platzky, L., & Walker, C. (1985). *The surplus people: Forced removals in South Africa*. Ravan Press.
Pottie, D. (2004). Local government and housing in South Africa: Managing demand and enabling markets. *Development in Practice, 14*(5), 606–618.
Raasch, S. (2001, August 31). Rural success stories: Small town folk doing it for themselves. *Dispatch*.
Raymond Mhlaba Local Municipality. (2018a). *Final integrated development plan*. Raymond Mhlaba Local Municipality.
Raymond Mhlaba Local Municipality. (2018b). *Spatial development framework-final draft*. Raymond Mhlaba Local Municipality.
Raymond Mhlaba Local Municipality. (2021). *Final annual report*. Raymond Mhlaba Local Municipality.
Raymond Mhlaba Local Municipality. (2022). *Draft integrated development plan*. Raymond Mhlaba Local Municipality.
Redding, S. (1992). South African blacks in a small town setting: The ironies of control in Umtata, 1878–1955. *Canadian Journal of African Studies, 26*(1), 70–90.
Reinders, S. (2020, November 15). *Why South Africans built an illegal settlement called Covid*. NPR.
Sapa. (2013, August 25). *ANC Free State expresses annoyance with shoddy contract work*. Mail & Guardian.
Satterthwaite, D., & Tacoli, C. (2006). The role of small and intermediate urban centres in regional and rural development. In C. Tacoli (Ed.), *The Earthscan reader in rural-urban linkages* (pp. 155–183). Earthscan.
Shackleton, M., & Blair, A. (2013). Perceptions and use of public green space is influenced by its relative abundance in two small towns in South Africa. *Landscape and Urban Planning, 113*, 104–112.
Skobba, K., Osinubi, A., & Tinsley, K. L. (2020). What about rural blight? Housing and neighborhood conditions in Southeastern small towns. *Rural Sociology, 85*(1), 85–110.
StatsSA (Statistics South Africa). (2013). *Community survey*. Statistics South Africa.
StatsSA (Statistics South Africa). (2016). *Community survey*. Statistics South Africa.
StatsSA (Statistics South Africa). (2021). *General households survey*. Statistics South Africa.
Thakrar, J. S. (2018). University-community engagement as place-making? A case of the University of Fort Hare and Alice. *Development Southern Africa, 35*(5), 673–688.

Van Niekerk, J., & Marais, L. (2008). Public policy and small towns in arid South Africa: The case of Philippolis. *Urban Forum, 19*, 363–380.

Westaway, A. (2012). Rural poverty in the Eastern Cape province: Legacy of apartheid or consequences of contemporary segregationism? *Development Southern Africa, 29*(1), 115–125.

White R. (2004). *Small town South Africa: The historical geography of selected Eastern Cape towns and current development initiatives within them* [Master's theses]. Rhodes University.

Xusa, P. H. L. (2005). Renewal of small town economies: The case of Alice, Eastern Cape Province, South Africa. *Africa Insight, 35*(4), 90–96.

Chapter 7
The Socio-economic Wellbeing of Small Mining Towns in the Northern Cape

Avril Gardiner and Ronnie Donaldson

7.1 Introduction

South Africa is renowned for its vast natural resources, and its economic prosperity is closely linked to the extraction and exportation of these raw materials. Historically, the country's economy has been driven by primary production sectors, particularly mining. The discovery of diamonds and precious metals such as gold and platinum played a significant role in the establishment and rapid development of many major South African towns and cities, with Johannesburg and Kimberley being prime examples (Spocter, 2021). The mining industry, which covers metal and mineral extraction up to the refining and smelting stages, is a crucial part of the resource sector, and many small mining towns are exposed to fluctuations in global resource markets. Given the constant variations in global demand for minerals and raw materials, the South African mining industry has gone through many cycles and will undoubtedly continue to do so.

In their case study on the impact of minerals on the regional development of the Northern Cape, Hanekom (1976) cites various parties' statements concerning the Northern Cape's development in the 1970s. The mayor of Springbok, welcoming delegates to a regional development conference in 1968, commented that Namaqualand would take the place of the Witwatersrand gold mines and Springbok would replace Johannesburg as South Africa's resource capital (Hanekom, 1976). Similar sentiments were echoed in a 1972 review article in the Cape Argus declaring that the Northern Cape would develop into a second Witwatersrand (Hanekom, 1976). More than 50 years later, we can safely say that these forecasts were overly bullish and optimistic. In fact, some of the towns may be in the same state of development as they were then, at best. The

A. Gardiner · R. Donaldson (✉)
Department of Geography and Environmental Studies, Small Town Research Unit, Stellenbosch University, Stellenbosch, South Africa
e-mail: rdonaldson@sun.ac.za

© The Author(s), under exclusive license to Springer Nature Switzerland AG 2023
R. Donaldson (ed.), *Socio-Spatial Small Town Dynamics in South Africa*, GeoJournal Library, https://doi.org/10.1007/978-3-031-37142-4_7

Northern Cape is the largest province in South Africa in terms of area (but with the smallest population), covering almost 30% of South Africa. Mining contributes about 24% to the Northern Cape economy and makes up nearly 7% of the country's total mining value, with significant deposits of iron ore, manganese, zinc, copper, lead, titanium, pig iron, zircon, and gypsum (https://www.nceda.co.za/index.php).

The spatial, social, and economic impact of mining activities on communities in South African towns and cities has long been the subject of scholarly investigation (for example, Hanekom, 1976; Hamman & Kapelus, 2004; Marais & Cloete, 2009; Rogerson, 2012; Marais, 2013; Burger et al., 2018; Marais et al., 2018). The recent increase in papers on mining towns can be attributed to several projects undertaken by Lochner Marais and colleagues on mining towns in the country (Marais, 2023; Sesele & Marais, 2022; Marais et al., 2022a, b; Marais et al., 2021; Burger et al., 2018; Marais, 2018; Marais & De Lange, 2021). However, one area of research that is lacking is a focus on human well-being in mining towns in South Africa.

Wellbeing can be defined as the benefit that all individuals in a society experience, which includes achieving adequate economic development (the objective dimension of wellbeing) and the resulting positive perception of people towards their proper place in society, i.e. the quality of life (the subjective dimension of wellbeing) (Ivkovic et al., 2014). The OECD's wellbeing index, developed in 2011 and updated in 2020, allows for the comparison of wellbeing across different geographic contexts. This index comprises two broad sub-sets: quality of life, which uses indicators such as health status, work and life balance, education and skills, material living conditions, civic engagement and governance, environmental quality, personal security, and subjective wellbeing, and material living conditions, which includes income and wealth, jobs and earnings, and housing (OECD, 2011, 2020). While many studies view wellbeing as a unitary phenomenon, most comparisons are made within industries, such as mining, rather than between industries, like forestry versus agriculture. Factors affecting wellbeing will vary not only within particular industries but also according to the region, historical period, type of resource being mined, and the indicator chosen to represent the state of wellbeing (Stedman et al., 2004).

In this chapter, we focus on a single industry comparison by selecting twenty-one mining small towns in the Northern Cape, identified from the provincial spatial development framework (Office of the Premier of the Northern Cape, 2012), as case study sites. We narrow down the analysis of census data for 2001 and 2011 (the only data available) to determine and compare the wellbeing rankings of these towns, given the availability and suitability of the census data.

7.2 Mining Towns: From Boomtowns to Resource Curse to Social Responsibility

The scholarly debate on mining towns can be grouped into three broad themes. The first theme is the concept of boomtowns, which explains why towns experience rapid growth when natural resources are discovered. The second theme investigates the impacts that resource wealth has on small mining towns, including the concepts of resource dependence and resource curse. The final theme focuses on the reality of deindustrialisation and downsizing of these mines and small towns and how mining companies deal with this, particularly through corporate social responsibility. These themes are all relevant to the impact on human well-being, which is affected by a range of social processes.

Research on boomtowns dates back to the 1970s, with most literature showing that rising commodity prices of the resource being mined affect the size of the population within the town and its economic activity (Lawrie et al., 2011). The first paper on the social implications of resource development in rural towns, known as the "Gillette syndrome," identified a range of social problems that occur in boomtowns, including dramatic increases in divorce, depression, school dropout rates, attempted suicides, juvenile delinquency, criminal activity, mental disorders, and other social problems. These findings sparked a keen interest among social scientists to investigate the boomtown phenomenon further.

Research has found that in most rural towns, social problems become apparent because of the collapse of informal social structures and the absence of new formal institutions to provide adequately for residents' social integration, social control, and well-being. Studies have been undertaken to assess the social impacts of the resource boom on rural communities. For example, Marais and Cloete (2009) used the extraordinary increase in the demand for housing in Kathu due to the increase in iron-ore mining since 2001 as a case study to illustrate the impact of mining activities on communities. Although their study concentrated on housing provision, it highlights certain key factors that should be considered when looking at the boom-bust cycle of mining towns in semi-arid and arid regions. The specific socio-economic features include variability, sparse population, low productivity, and remoteness. The authors concluded that in terms of many development indicators, these areas lag behind their counterparts in areas with better habitable conditions.

Resource dependency refers to the economic reliance of people living in resource-dependent towns on the extraction of natural resources (Freudenburg, 1992; Freudenburg & Wilson, 2002). This phenomenon is particularly relevant to mining towns, where the closure of a mine due to resource depletion can leave the community, or most of those employed in the mining industry, without a source of income. Black et al. (2005) research shows that the effects of a mining bust extend beyond the mining sector, with job losses in construction, retail, and services sectors. The natural life cycle of a small town that participates in the mining sector can lead to the phenomenon of the resource curse, whereby natural resource abundance fails to translate into positive economic and social development. Atkinson and Hamilton (2003) explain the resource curse as a negative relationship between natural resources and the growth rate of per capita gross domestic product. Scholars have attempted to explain this phenomenon and develop strategies to avoid it, but the failure of policy has been central to many of these efforts. Countries where governments consume resource revenues experience the effects of the resource curse more severely than countries that reinvest profits in a sustainable manner (Atkinson & Hamilton, 2003). The closure of a mine differs from that of an industrial enterprise, as it typically affects smaller communities located in remote areas. Despite the anticipated nature of mine closures, research shows that they are not always perceived as such. During a mining boom, crime often increases, but research has also shown that the opposite can occur, as demonstrated by the case study of the gold-mining town of Matjhabeng (formerly Welkom), which also disproportionately affects women (Sesele & Marais, 2022). Crucial planning and policy-making questions include whether communities have a future in towns after mining and whether sustainable methods can be implemented to alleviate the effects of the resource curse. In his pioneering work, Marais (2023) questions South Africa's comprehensive housing programs linked to homeownership, which promote place attachment, stability, and wealth among mine workers. He argues that such policies create long-term dependencies that are difficult to manage when a mine closes and challenges the blind application of these policies during boom periods.

Corporate social responsibility (CSR) has been increasingly promoted in mining towns since the 1990s to address the negative reputation of mining companies for causing environmental devastation and social disruption through community displacements (Littlewood, 2004; Dashwood, 2012). When a mine closes or downsizes, many community members leave the town, and the loss of social networks can exacerbate the stress of unemployment, especially in well-established mining communities with limited external connections. Mining companies implement CSR using three general models: independent funding, employing non-profit organizations, or retaining CSR within their own company structures. However, the basis for deciding which model to use and the factors that determine CSR performance are not clear (Kapelus, 2002). In South Africa, legislation passed in 2002 requires private and public entities to plan for sustainable community development. CSR is linked to planning phases that include pre-mining impact assessments, regular community monitoring, planning for viable economic activities, utilizing mine infrastructure, providing land for alternative uses, carrying out environmental planning, and aligning mining company objectives with those of the government (Warhurst & Noronha, 2000; Marais, 2013). Marais (2013) notes that a long history of role-players in the development of mines has influenced the development path in the Free State Goldfields, going through five phases with different outcomes. However, many of the social responsibility plans were ineffective due to unrealistic growth predictions, lack of cooperation between local government and mining companies, and short-term, unsustainable projects that do not lead to long-term sustainable effects. Marais (2013) argues that most mining companies see social and labor plan projects as a way to comply with the law and invest in short-term, unsustainable projects that provide immediate rather than long-term sustainable effects. Therefore, there is a need for mining companies to adopt sustainable CSR practices that align with the needs of the community and prioritize long-term sustainable effects.

A common question is whether mining communities can maintain a positive outlook after a bust cycle in a resource-dependent town. Brown et al. (2005) addressed this question by examining changes in community satisfaction over time. They concluded that residents' subjective interactions with their communities tend to improve over time. Furthermore, there is an inverse relationship between community satisfaction and population growth, which is consistent with prior research suggesting that rapid population growth and large influxes of newcomers can lead to a decline in social integration and community satisfaction. The greatest increases and decreases in community satisfaction occurred during time periods when population growth was relatively stable, indicating that community cohesion is not solely determined by population fluctuations but rather by a multitude of factors. Hajkowicz et al. (2011) investigated the connection between social well-being and residents of mining towns. They found that the type of mineral being mined also correlates with the well-being levels of communities. The label "resource dependence" may obscure a range of variables in the relationship between specific resource industries, regions, and indicators of well-being (Stedman et al., 2004). Freudenburg and Wilson (2002) argue that a correlation exists between the type of mineral being mined and the socio-economic status of people. In the United States, areas that extracted minerals for energy during the 1970s and 1980s fared much better in socio-economic terms than areas that extracted iron, where people were living in poverty, despite both resource sectors being in the boom phase of their cycle. Although the Northern Cape is not necessarily rich in gold resources, it is similar to the recent decline in job opportunities in Orkney and Klerksdorp due to the drop in levels of gold production, which demonstrates this dependency relationship (Kleynhans, 2012).

7.3 Conceptual Context and Methods

Studies that measure the well-being of communities use a wide range of indicators, such as income, consumption, residential land, wages, rents, local amenities, natural environment, and environmental pollution, among others (Smith et al., 2001; Tonts et al., 2011). According to the OECD (2011:4), progress can be defined simply as "improvements in the well-being of people and households." This includes analyzing the diverse experiences and living conditions of individuals, as well as the functioning of the economic system. To measure well-being, the OECD has developed a "better life index." The framework is based on three conceptual pillars: material conditions, quality of life, and sustainability. Each of these pillars has relevant sub-dimensions.

The first of the three pillars is material conditions, which refers to the command over physical commodities. Measuring this phenomenon requires analyzing people's income, assets, and consumption, as well as how these economic resources are distributed among different individuals and population groups (OECD, 2012).

The second pillar is quality of life, which focuses on indicators other than those related to material conditions. While economic resources are important, they are not the only factors that determine people's well-being. These quality of life indicators include education, human contact, health status, environmental quality, jobs, governance, security, and civic engagement. Measuring people's subjective experiences is also important in assessing the levels of well-being in a community. Thus, measuring quality of life involves considering economic and non-economic, subjective and objective indicators, and disparities and similarities between different population groups (OECD, 2012).

The third pillar is sustainability, which poses many challenges when assessing over time. Future well-being cannot be predicted in the present due to changes in taste and day-to-day advancements in technology (OECD, 2012). However, it is possible to measure the current levels of resources that shape well-being in communities and monitor how these resources are being managed for future generations. Sustainability can therefore be measured in the following four resources or "capitals" that will affect future well-being: economic, natural, human, and social capitals. While the levels of these resources may not be the only determinants of well-being over time, they offer a quantitative means of measuring and examining the link between the present situation and the future.

An aggregation of indicators is referred to as an index or composite indicator. Composite indicators are constructed using sub-indicators that have no common, meaningful unit of measurement, and no obvious way to weight these sub-indicators. Composite indicators have both advantages and disadvantages (see Saisana, 2004). In this study, an equal weighting method (Nardo et al., 2005) was used to weigh indicators within each of the two domains of the OECD well-being indicator framework, namely, quality of life and material living conditions (OECD, 2011, 2020). The importance of the four dimensions for well-being is summarized in Table 7.1.

In many composite indicators, all variables are given equal weight when there are no statistical or empirical grounds for choosing a different scheme. The rationale for this is that the researcher recognizes the equal importance of all material living conditions, rather than insufficient knowledge of the relationships between indicators or which model to apply. The impact of equal weighting on the composite indicator also depends on whether equal weight is applied to single indicators or components. Census data at the small-place level was extracted and accessed through SuperCROSS (the official census interactive data program), and then analyzed in Excel and STATISTICA. For data analysis, standard reliability and item analyses were conducted using t-tests. A box and whisker plot were used as a means of summarizing data measured on an interval scale. The data analyses aimed to demonstrate any changes in social well-being over the ten-year period and rank each town according to its level of well-being.

There are a total of 22 small towns classified as mining towns in the Northern Cape (as shown in Fig. 7.1 and Table 7.2). These towns are clustered in three regions of the Northern Cape. The first cluster consists of towns on the West Coast, which are known for their offshore diamond mining operations. The second cluster, comprising Carolusberg, Concordia and Okiep, is located just east of the West Coast and is historically known for copper mining. These towns were particularly affluent when the copper price was very high but went into decline when the demand for copper fell worldwide. The third cluster is located near the northeastern border of the Northern Cape, close to Kimberley, and includes Barkly West, Kathu, Sishen, Blackrock, Hotazel, Postmasburg, Danielskuil, Delportshoop, Ulco, and Lime Acres. These towns are known for open-mine diamond mining and, more recently, iron ore, manganese, and limestone mining. Table 7.2 provides a summary of the

Table 7.1 Two domains and four dimensions for OECD well-being indicators

Two domains	Dimensions	Why do they matter for well-being?	Data used from census 2001 and 2011
Material living conditions	Income and wealth	Income and wealth are vital factors that contribute to the well-being of individuals and societies. They provide people with the means to meet their needs, expanding their consumption possibilities. Additionally, wealth enables individuals to manage their consumption over time and protect themselves from unforeseen shocks that may result in poverty and destitution. In addition to economic advantages, income and wealth also bring about non-economic benefits such as improved health, access to education, higher levels of life satisfaction, and the opportunity to live in safer and cleaner areas.	% of the households earning less than R76 400 per annum % of individuals with monthly income lower than R1600 per month
	Jobs and earnings	Both job availability and earnings are crucial factors that contribute to well-being. They not only increase people's access to resources but also provide opportunities to fulfil personal ambitions, develop skills, feel useful in society, and build self-esteem. Societies with high levels of employment tend to be more prosperous, politically stable, and healthier. On the other hand, unemployment has a significant negative impact on subjective well-being, with effects that are much greater than the income loss associated with it. Furthermore, evidence suggests that this impact can persist over time and that psychological resilience to unemployment is low.	% of the population that is employed % of the labour force that work in the formal sector
	Housing	Having adequate accommodation is at the forefront of the hierarchy of human material needs. Housing represents the largest component of many households' expenses and is essential for meeting basic needs. Poor housing conditions can also have a significant impact on people's health, both mentally and physically, as well as their family functioning, including household member relationships and child development. Additionally, it can impact the ability to conduct basic social activities, such as inviting guests over.	% of households living in informal housing % of the households that owning the dwelling (which is fully paid off) % of households with no access to water % of households with access to flushing toilets % of households with more than five people living in the dwelling

(continued)

Table 7.1 (continued)

Two domains	Dimensions	Why do they matter for well-being?	Data used from census 2001 and 2011
Quality of life	Education and skills	Education is a fundamental need and an important aspiration for individuals, with a strong influence on their well-being. Individuals with higher levels of education typically earn higher wages and have a greater likelihood of securing employment. They also tend to live longer, report better health, and experience fewer chronic diseases and disabilities. Furthermore, better-educated individuals are more engaged in politics and their local community, commit fewer crimes, and rely less on social assistance. At the societal level, improved education is associated with higher GDP growth, increased tax revenues, and reduced social expenditures.	% of the population over the age of 20 years with an educational qualification lower than matric % of the population with some form of tertiary education

Source: compiled from OECD (2011)

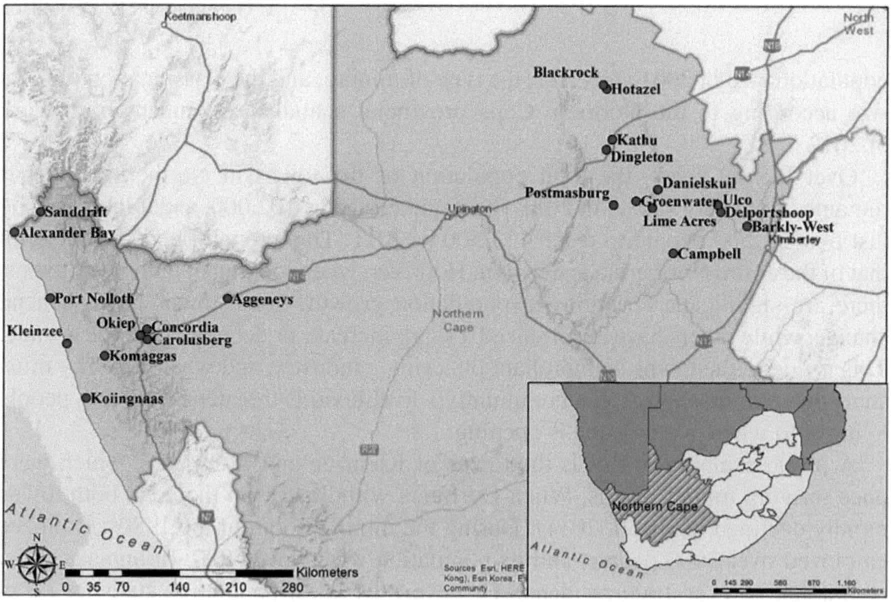

Fig. 7.1 Location of small mining towns in the Northern Cape (Source: Authors, 2022)

Table 7.2 Summary of mining towns in the Northern Cape

Town name	Population 2001	Population 2011	Type of mining	Size
Aggeneys	2057	2262	Zinc/Lead/copper/silver	Medium
Alexander Bay	1452	1736	Diamonds	Small
Barclay West	7101	8243	Diamonds	Large
Blackrock	1002	403	Manganese	Small
Campbell	1768	2175	Copper	Medium
Carolusberg	968	1335	Copper	Small
Concordia	4036	4988	Copper Granite	Medium
Danielskuil	6730	13,597	Asbestos/Diamonds/Marble	Large
Delpoortshoop	2831	4787	Diamonds	Large
Dingleton/Sishen	2866	11,084	Iron Ore	Medium
Groenwater	517	739		Very Small
Hotazel	1232	1756	Manganese	Small
Kathu	8,247,728	11,510	Iron Ore	Large
Kleinzee	2712	728	Diamonds	Medium
Koingnaas	724	105	Diamonds	Medium
Komaggas	3315	3116	Diamonds	Medium
Lime Acres	3729	4406	Lime Stone/Diamonds	Medium
Okiep	5241	6299	Copper	Medium
Port Nolloth	4655	6085	Diamonds (shipping)	Medium
Postmasburg	13,962	30,085	Manganese	Large
Sanddrift	1146	1854	Diamonds	Small
Ulco	1063	860	Lime Stone/Cement	Small

population size in 2001 and 2011, the type of mining, and the classification of town size according to the Northern Cape provincial spatial development framework of 2012.

Over the 10 years, the total population of the towns in Northern Cape has remained relatively stable. In 2001, the population was 102,000, and it increased by just over 11,500 people to reach 113,500 in 2011. This growth rate is in line with that of the entire Northern Cape region. However, when looking at individual towns, there are significant variations in population growth. Some towns have seen no change, while others have experienced a sharp increase or decrease over the decade. This is due to the towns being reliant on a single industry, and when a nearby mine shuts down or downsizes, the community's livelihood is threatened, forcing people to move to where a new mine is opening.

A prime example of this is the towns of Kleinzee and Koingnaas, which were once thriving mining towns. When De Beers withdrew from the area, both towns rapidly declined (Stilwell, 2011). During the mining boom in the 1980s, Kleinzee employed over 3000 people, and its population was above 6000. Koingnaas had a population of over 1000 residents. However, the latest population statistics show that the population of Kleinzee has decreased from 2712 in 2001 to 728 in 2011, and Koingnaas' population has fallen from 724 to 105 in the same period.

On the other hand, some towns have experienced massive population growth in the ten-year period between 2001 and 2011. These include Kathu, Sishen, Postmasburg, Danielskuil, and Delportshoop. Kathu and Sishen are located close to each other, as are Postmasburg and Danielskuil. This influx of people was due to the booming iron ore and manganese mining industry in the area. Large companies such as Kumba, Assmang, and Indwala moved into the region and transformed the economic landscape through their extensive operations. People moved to these towns, hoping for a better life, as billions of Rands were generated from the export of these natural resources. The populations of other mining towns in the area have remained more or less stable. The stable population in some towns can be attributed to various factors, including the shift towards other industries like agriculture, as is the case in Barkly West. Another reason is that although some mines have downsized or closed before 2001, there are still mining operations in the area, albeit on a smaller scale. For instance, people in Alexander Bay and Port Nolloth are still employed in offshore diamond mining, and the populations of Ulco and Lime Acres have remained stable due to the consistent limestone and cement markets and the mines' continued operation. Similarly, the populations of Concordia and Aggeneys, where there is still copper mining on a smaller scale, have remained stable. However, the population of Blackrock has shown a sharp decline in its male population, which can be attributed to more women moving into the town due to long-term job opportunities, and the wives of male workers moving into the town after the new mine became operational. The male population in Blackrock declined from just above 70% to a median of about 50%, while the female population grew from being just below 30% in 2001 to almost 50% in 2011. Although the increase in males in the small mining towns in the Northern Cape may seem small, it is significant since mining is still a very male-dominated industry.

7.4 Material Living Conditions

For the domain of material living conditions, three dimensions were considered: income and wealth, jobs and earnings, and housing.

7.4.1 Income and Wealth

This dimension reflects the level of economic stability in communities in the vicinity of mining operations. Table 7.3 ranks the towns according to the percentage of households earning less than R19,600 per annum. The worst-performing towns are highlighted in bold; the best-performing towns are highlighted in italic. Earning less than R19,600 can be classified as extremely poor, with households living on just over R1000 per month. The town's ranking is higher if it has fewer low-income households. Table 7.3 also ranks individual monthly incomes and shows that in almost all cases, almost 50% of the people earn either no income or less than R1600 per month.

Table 7.3 Ranking of household income below R19 600 per annum

	Household income below R19 600 per annum				Ranking of individual monthly income below R1600 per month			
	2001		2011		2001		2011	
Town	%	Rank	%	Rank	%	Rank	%	Rank
Aggeneys	15.8	3	*5.76*	*1*	54.22	3	45.49	5
Alexander bay	16.4	4	11.11	2	56.51	5	54.84	10
Barkly West	64.9	19	43.74	19	85.81	17	76.16	16
Blackrock	50.4	12	28	10	59.32	7	40.74	3
Campbell	84.9	21	51.65	21	**95.41**	**22**	80.14	18
Concordia	62.7	17	36.34	15	87.63	19	77.39	17
Carolusberg	39.1	9	21.8	9	76.72	11	54.93	11
Danielskuil	68	20	39.6	17	88.34	20	81.66	20
Delportshoop	63.8	18	46.35	20	87.51	18	83.08	21
Sishen	56.2	14	28.49	12	82.95	14	52.84	9
Groenwater	**90.9**	**22**	**58.57**	**22**	95.02	21	80.89	19
Hotazel	42.5	10	14.07	5	67.83	8	48.29	7
Kathu	31.6	6	13.79	3	55.27	4	41.2	4
Kleinzee	14	2	14	4	29.04	2	*28.51*	*1*
Koingnaas	*5.17*	*1*	15	6	21.7	1	40	2
Komaggas	43.4	11	43.46	18	77.07	12	**86.9**	**22**
Lime Acres	28.3	5	20.34	8	57.38	6	46.97	6
Okiep	55.6	13	37.3	16	84.83	16	73.94	15
Port Nolloth	57.7	15	36.29	14	83.18	15	70.61	14
Postmansburg	57.9	16	32.1	13	82.89	13	67.68	13
Sanddrif	36.3	8	28.24	11	70.19	10	62.62	12
Ulco	31.8	7	17.71	7	68.47	9	50.7	8

7.4.2 *Jobs and Earnings*

This indicator uses the level of employment as a component of the composite indicator index. The higher the percentage of people employed in the town, the better its ranking (see Table 7.4). Although Koingnaas and Kleinzee are ranked at the top, it should be noted that the populations of these towns are very small and mining in the vicinity is almost non-existent at the moment. Therefore, the people who still live in these towns are employed in other sectors. Groenwater has the lowest ranking, with only 9% of the population employed.

In the 2001 census, information about individuals' work sectors, such as mining and quarrying, agriculture, and manufacturing, was recorded. However, in the 2011 census, only the differentiation between working in the formal or informal sector was noted. Thus, the ranking of towns was determined based on the percentage of workers in the formal sector (as shown in Table 7.4). It is noteworthy that the number of people employed in the formal sector has declined in these towns. This indicates that significant downsizing and mine closures have altered the job market, and more people have had to resort to the informal economic sector to make a living.

Table 7.4 Ranking according to population employed and according to those working in the formal working sector

	Ranking according to population employed				Ranking according to those working in the formal sector			
	2001		2011		2001		2011	
Town	%	Rank	%	Rank	%	Rank	%	Rank
Aggeneys	62.31	6	44.16	11	53.17	6	40.05	6
Alexander Bay	58.33	8	46.29	8	51.17	7	44.39	3
Barkly West	28.69	16	17.22	18	24.88	16	13.94	18
Blackrock	79.02	3	52.24	2	69.7	3	36.57	11
Campbell	**11.46**	**22**	12.69	21	**10.32**	**22**	9.66	21
Concordia	28.42	17	21.47	16	24.03	17	19.3	15
Carolusberg	48.32	10	45.39	9	45.69	10	38.2	8
Danielskuil	26.81	19	20.34	17	22.47	19	16.11	17
Delportshoop	21.6	20	14.85	20	19.24	20	10.4	19
Sishen	26.93	18	44.64	10	23.73	18	34	12
Groenwater	12.94	21	**8.54**	**22**	11.95	21	**6.91**	**22**
Hotazel	60.71	7	47.86	7	50.85	8	38.29	7
Kathu	65.06	4	51.41	3	55.21	4	45.15	2
Kleinzee	86.03	2	50.41	4	79.07	2	42.56	4
Koingnaas	*88.83*	*1*	*68.57*	*1*	*81.81*	*1*	*60*	*1*
Komaggas	40.03	12	15.98	19	32.5	12	10.11	20
Lime Acres	63.55	5	49.18	5	54.49	5	37.67	9
Okiep	32.98	15	26.35	15	29.36	14	18.01	16
Port Nolloth	37	13	35.7	13	32.02	13	28.65	13
Postmansburg	34.04	14	30.41	14	27.22	15	19.78	14
Sanddrif	47.66	11	40.94	12	43	11	37.38	10
Ulco	54.74	9	48.78	6	47.15	9	40.42	5

7.4.3 Overall Ranking

Resource-dependent regions and localities are characterised by unique and specific circumstances, including reliance on particular industries to drive the local economy, dependence on non-renewable resources, vulnerability to global markets and resource prices, and limited alternative economic opportunities (Halseth, 1999). Tonts (2010) argues that although these industries have the potential to generate significant economic returns, their regional employment outcomes often prove unsustainable, failing to create long-term opportunities for local residents. Using the OECD's (2011) Better Life Index as a guideline, three indicators were grouped together to measure economic well-being in communities using 2001 and 2011 census data. These indicators include the percentage of households earning less than R76 400 per annum, the percentage of individuals with monthly incomes below R1600, the percentage of the population that is employed, and the percentage of the labour force working in the formal sector. The four rankings were then added up, and the result was divided by four to depict economic well-being in each town (Table 7.5).

Table 7.5 Overall ranking of economic dimension

	2001		2011	
Town	Average	Rank	Average	Rank
Aggeneys	4.5	3	5.75	4
Alexander Bay	6	6	5.75	4
Barkly West	17	17	17.75	17
Blackrock	6.25	7	6.5	6
Campbell	**21.75**	**22**	20.25	21
Concordia	17.5	18	15.75	16
Carolusberg	10	10	9.25	10
Danielskuil	19.5	20	17.75	17
Delportshoop	19	19	20	20
Sishen	16	16	10.75	11
Groenwster	21.25	21	**21.25**	**22**
Hotazel	8.25	8	6.65	6
Kathu	4.5	3	3	2
Kleinzee	2	2	3.25	3
Koingnaas	1	1	2.5	1
Komaggas	*11.75*	*12*	*19.75*	*19*
Lime Acres	5.25	5	7	9
Okiep	14.5	14	15.5	15
Port Nolloth	14	13	13.5	13
Postmansburg	14.5	14	13.5	13
Sanddrif	10	10	11.25	12
Ulco	8.5	9	6.5	6

7.4.4 Housing and Infrastructure

In this section, the level of household infrastructure in mining towns in the Northern Cape is measured and ranked according to indicators that form part of a composite indicator showing overall quality of life. Five indicators are used to indicate the levels of household infrastructure: the percentage of households living in informal housing, the percentage of households that own their dwelling (which is fully paid off), the percentage of households with no access to water, the percentage of households with access to flushing toilets, and the percentage of households with more than five people living in the dwelling. Table 7.6 displays the ranking of towns based on the percentage of households living in informal housing. A higher ranking indicates a lower percentage of households living in informal housing. Although this ranking has a large variance, it becomes more balanced when considered alongside the other indicators. The towns either have a very low number of people living in informal housing or a substantially higher number, with no middle ground. Some towns have less than 1% living in informal housing but rank near the middle and not at the top (as in Okiep).

Table 7.6 Ranking according to informal housing and owned (fully paid off) dwelling ranking

	Rank according to informal housing				owned (fully paid off) dwelling ranking			
	2001		2011		2001		2011	
Town	%	Rank	%	Rank	%	Rank	%	Rank
Aggeneys	*0*	*1*	1	11	0.45	20	0.52	18
Alexander Bay	3.05	11	*0*	*1*	2.34	16	**0**	**19**
Barkly West	33.92	21	12.02	16	57.25	6	45.48	8
Blackrock	1.87	8	*0*	*1*	1.87	17	**0**	**19**
Campbell	4.76	12	26.92	20	71.45	4	20.88	12
Concordia	2.31	10	0.21	5	84.96	3	69.12	4
Carolusberg	2.25	9	1.5	10	52.96	8	63.91	5
Danielskuil	16.72	18	15.05	17	53.23	7	43.12	9
Delportshoop	7.93	14	11.98	15	51.22	9	51.43	7
Sishen	6.15	13	**26.2**	**22**	47.38	11	15.55	14
Groenwater	**86.23**	**22**	24.29	19	9.13	13	*82.86*	*1*
Hotazel	1.78	7	9.5	14	0.6	19	1.51	17
Kathu	1.42	5	1.89	12	7.31	14	10.1	15
Kleinzee	0.19	2	*0*	*1*	0.39	21	**0**	**19**
Koingnaas	0.7	4	5	13	0.7	18	20	13
Komaggas	8.59	15	0.71	6	*90.7*	*1*	79.51	2
Lime Acres	0.45	3	0.85	7	5.15	15	7.18	16
Okiep	10.59	17	0.88	8	89.02	2	71.28	3
Port Nolloth	9.44	16	1.15	9	63.61	5	56.49	6
Postmansburg	16.84	19	29.11	21	49.55	10	42.46	10
Sanddrif	28.48	20	21.64	18	34.14	12	24.12	11
Ulco	1.6	6	*0*	*1*	**0**	**22**	**0**	**19**

Table 7.6 displays the ranking of tenure status for dwellings. Groenwater is ranked at the top, with 86% of households owning their dwellings. This is the only indicator in which Groenwater ranks at the top. It is possible that government housing projects have contributed to this result, as the houses are given to the permanent population of this very small town. Additionally, this indicator can tie in with economic stability. When more people own the houses they live in, they tend to feel more financially secure and stable, especially if the house is fully paid off.

Household infrastructure is included in this composite index to indicate the level of amenities within the house itself. Three datasets were used for this indicator: flushing toilets, electricity for home lighting, and access to water connected to the house. This data demonstrates that there are sufficient facilities to improve people's health. Flushing toilets in the datasets refer to toilets connected to the sewerage system or a septic tank. Having access to a flushing toilet and clean running water reduces health risks for households compared to those without this basic household infrastructure. Table 7.7 displays the ranking of towns based on these household infrastructure indicators.

7.4.5 *Overall Housing Rank*

There were no significant improvements in rank (Table 7.8) for any of the towns, except for Blackrock, which moved from tenth to second place in the ranking. This change may be attributed to the 60% decline in population, from 1000 to 400 people, resulting in a higher percentage of a smaller population having access to these infrastructure developments. Interestingly, Kathu ranks quite low in tenth place, despite ranking in the top five for the other indicators. This may be because Kathu has a high number of people who rent, which contributes to a low ranking in the ownership of houses indicator.

7.5 Quality of Life

For the domain of quality of life, the domain of education and skills were considered.

7.5.1 *Education and Skills*

Shao and Yang (2014) note that human capital plays a vital role in driving economic growth and reversing the effects of the resource curse. They argue that the resource curse has a crowding-out effect on human capital development, particularly education. In resource-based economies, people often view investment in human capital as unlikely to yield expected returns, leading to a decline in the desire for education and

Table 7.7 Rankings of household infrastructure (water, toilets, household size, electricity)

Town	H20 Connected to house				Flushing toilets in home				HH size5+		Electricity in home			
	2001		2011		2001		2011		2011		2001		2011	
	%	Rank	%	Rank	%	Rank	%	Rank	%	Rank	%	Rank	%	Rank
Aggeneys	95.07	4	98.43	2	97.76	5	99.84	4	23.04	14	*100*	*1*	*100*	*1*
Alexander Bay	81.31	7	97.08	4	97.71	6	98.25	5	20	11	*100*	*1*	*100*	*1*
Barkly West	52.24	14	60.54	15	54.18	17	86.79	13	31.1	20	65.94	21	82.97	20
Blackrock	32.28	17	98	3	98.13	4	98	6	14	5	99.06	7	100	1
Campbell	**6.35**	**22**	**6.5**	**9**	**9.52**	**22**	**4.95**	**22**	**32.04**	**22**	83.71	17	**68.51**	**22**
Concordia	57.06	12	80.46	10	43.82	19	69.54	18	21.85	13	95.64	11	98.95	7
Carolusberg	76.75	9	79.1	11	87.54	11	78.36	15	18.05	9	98.87	8	92.48	14
Danielskuil	47.07	15	68.89	14	84.36	12	94.71	12	31.81	21	79.61	19	91.24	16
Delportshoop	54.86	13	43.9	18	70.68	15	84.42	14	29.87	19	89.63	15	85.94	19
Sishen	34.63	18	41.9	20	83.84	13	97.98	8	13.95	4	83.46	18	74.29	21
Groenwater	9.37	21	28.57	21	11.57	21	50	20	24.29	15	**9.37**	**22**	88.57	17
Hotazel	81.73	6	89.45	9	95.46	10	97.99	7	15	6	96.05	10	96	12
Kathu	91.06	5	92.92	6	98.4	3	97.92	9	16.71	8	99.34	6	97.64	10
Kleinzee	*98.66*	*1*	*100*	*1*	*99.81*	*1*	*100*	*1*	*3*	*1*	*100*	*1*	*100*	*1*
Koingnaas	95.11	3	95.24	5	99.3	2	*100*	*1*	4.76	2	*110*	*1*	*100*	*1*
Komaggas	25.51	20	58.66	16	21.42	20	53.36	19	25.8	18	94.02	12	98.59	8
Lime Acres	77.13	8	90.61	7	96.82	9	96.42	11	12.14	3	97.96	9	96.93	11
Okiep	65.22	10	75.48	13	60.4	16	71.28	17	25.61	17	78.91	20	94.92	13
Port Nolloth	64.04	11	78.16	12	97.4	8	97.37	10	18.23	10	94	13	98.03	9
Postmansburg	39.16	16	45.49	17	80.1	14	72.34	16	25.04	16	88.05	16	87.65	18
Sanddrif	34.35	19	43.86	19	53.84	18	41.52	21	21.05	12	90.18	14	92.4	15
Ulco	96.69	2	89.58	8	97.6	7	*100*	*1*	16.67	7	100	1	98.96	6

Table 7.8 Overall housing rankings

	2001		2011	
Town	Average	Rank	Average	Rank
Aggeneys	6.2	3	8.33	6
Alexander	8.2	6	6.83	4
Barkly Wes	15.8	20	15.33	17
Blackrock	10.6	10	5.83	2
Campbell	15.4	19	**20**	**22**
Concordia	11	12	9.5	9
Carolusberg	9	8	10.66	11
Danielskuil	14.2	16	14.83	15
Delportshoop	13.2	14	15.33	17
Sishen	14.6	17	14.83	15
Groenwater	**19.8**	**22**	15.5	19
Hotazel	10.4	9	10.83	12
Kathu	6.6	4	10	10
Kleinzee	*5.2*	*1*	*4*	*1*
Koingnaas	5.6	2	5.83	2
Komaggas	13.6	15	11.5	13
Lime Acres	8.8	7	9.17	7
Okiep	13	13	11.83	14
Port Nolloth	10.6	10	9.33	8
Postmansberg	15	18	16.33	21
Sanddrif	16.6	21	16	20
Ulco	7.6	5	7	5

a problem in the development of human capital (Shao & Yang, 2014). Stijns (2006), however, argues that the relationship between human capital and natural resource wealth is not as robust as previously believed. The rational allocation of resource rent in resource-rich countries is a prerequisite for promoting better human capital. Bravo-Ortega and De Gregorio (2007) empirically show that Scandinavian countries successfully evaded the resource curse by investing in higher levels of human capital, which compensated for and even eradicated negative effects brought on by resource booms.

Education has a significant impact on human well-being. It is a basic need and an essential part of life to which people aspire. Individuals with better education earn higher wages and have a higher probability of securing jobs (OECD, 2012). According to the Better Life Index (as shown in Table 7.1), people with better education live longer lives, have better health statuses, and lower levels of chronic diseases and disabilities (OECD, 2016). As a rule, the higher the level of education within a society, the higher the GDP growth and tax revenues, and the lower the social expenditure (OECD, 2012).

The composite indicator index (Table 7.9) portrays the dimension of education and skills using two indicators: the percentage of the population over the age of 20 with an educational qualification lower than matric, and the percentage of the

Table 7.9 Education and skills rankings

Town	Age over 20 with less than matric				Population with tertiary qualification				Overall ranking of education and skills			
	2001		2011		2001		2011		2001		2011	
	%	Rank	%	Rank	%	Rank	%	Rank	%	Rank	%	Rank
Aggeneys	60.91	6	59.54	4	5.48	6	5.93	5	6	6	4.5	4
Alexander Bay	64.64	7	80	9	4.95	9	1.61	16	8	8	12.5	11
Barkly West	77.63	12	87.9	14	3.4	11	2.42	13	11.5	11	13.5	13
Blackrock	76.97	10	79.31	8	5.37	8	8.47	4	9	9	6	6
Campbell	**92.4**	**22**	95.89	21	**0.34**	**21**	1.1	19	**21.5**	**22**	20	20
Concordia	82.86	19	92.03	17	1.51	19	1.9	14	19	19	15.5	17
Carolusberg	75.26	9	95.38	20	3.43	10	0.84	20	9.5	10	20	20
Danielskuil	82.12	18	90.26	16	2.23	15	2.91	12	16.5	17	14	14
Delportshoop	78.8	15	95.23	19	3.39	12	1.46	17	13.5	13	18	18
Sishen	78.75	14	77.55	7	1.59	17	**0.69**	**22**	15.5	16	14.5	16
Groenwater	91.54	21	**98.46**	**22**	0.59	20	0.77	21	20.5	20	**21.5**	**22**
Hotazel	66.25	8	56.47	3	5.47	7	11.39	2	7.5	7	2.5	2
Kathu	47.15	1	36.62	1	12.13	3	15.02	1	2	2	1	1
Kleinzee	56.99	2	43.53	2	*18.82*	*1*	5.29	8	*1.5*	*1*	5	5
Koingnaas	57.65	4	72.73	6	6.6	5	4.35	9	4.5	5	7.5	7
Komaggas	86.19	20	93.2	18	0.18	22	1.19	18	21	21	18	18
Lime Acres	57.06	3	67.53	5	7.96	4	9.94	3	3.5	3	4	3
Okiep	80.89	17	87.4	13	1.55	18	1.67	15	17.5	18	14	14
Port Nolloth	77.62	11	81.99	10	1.85	16	5.3	7	13.5	13	8.5	8
Postmansburg	79.12	16	88.46	15	2.24	14	3.76	10	15	15	12.5	11
Sanddriff	78.64	13	84.29	11	3.15	13	3.14	11	13	12	11	10
Ulco	60.45	5	84.55	12	17.2	2	5.41	6	3.5	3	9	9

population with some form of tertiary education. The data clearly shows that the workforce in all the towns is largely unskilled, except for Kathu, where only 36% of the population over the age of 20 did not finish matric. Kathu also ranks at the top in terms of people with some form of tertiary education, with 15% of the population. The next nearest is Hotazel with 11%. This indicator also shows that the levels of specialization are generally low in most cases.

7.5.2 Composite Indicator for Well-Being

The composite indicator index indicates that the level of socio-economic wellbeing is directly linked to the stage of the mining cycle in which the town finds itself. Towns that are still viable and profitable have a higher level of wellbeing, whereas those that have experienced the boom-bust cycle are near the bottom of the ranking, except for Kleinzee and Koingnaas. These two towns (Table 7.10), which are ranked 1st and 2nd respectively, have become almost ghost towns due to a sharp decline in their populations to just over 700 in Kleinzee and 105 people in Koingnaas. This

Table 7.10 Overall ranking of towns according to human wellbeing and quality of life composite indicator index

	Overall Rank	
Town	2001	2011
Aggeneys	4	4
Alexander Bay	7	6
Barkly West	16	18
Blackrock	9	4
Campbell	**21**	**21**
Concordia	18	12
Carolusberg	10	11
Danielskuil	20	17
Delportshoop	14	20
Sishen	18	12
Groenwater	**21**	**21**
Hotazel	8	8
Kathu	3	3
Kleinzee	*1*	*1*
Koingnaas	2	2
Komaggas	16	19
Lime Acres	5	6
Okiep	13	15
Port Nolloth	11	10
Postmansburg	15	16
Sanddrif	12	12
Ulco	6	8

suggests that the remaining residents have sufficient resources to maintain a high standard of living and wellbeing. Furthermore, the departure of mining operations from the area has resulted in a decrease in the need for further development, as the relatively high levels of wellbeing and quality of life for the remaining residents do not require significant improvement. Interestingly, Campbell and Groenwater achieved the same average score and ranking.

7.6 Conclusion

The demographics of the small mining towns have not changed significantly over the 10-year census period between 2001 and 2011. While population growth has occurred, in most cases it was not more or less than the average for South Africa. The exceptions were towns which experienced (i) significant growth in the form of new mining operations opening or (ii) decline, where operations were downscaled or closed. What is apparent is that in these single-industry towns, population figures can be directly linked to the stage of the mining cycle. Due to the lack of other industries and sectors to absorb the over-supply of workforce (e.g. Barkly West, Kleinzee and Koingnaas), people move to areas with greater demand (e.g. Postmasburg). The gender and race profiles suggest that white males in these towns have moved away due to mining operations downsizing and mechanisation reducing the demand for specialised and skilled labour.

Through data analysis, it can be confirmed that the government has fulfilled its promise of providing the communities with basic services and infrastructure, with many people living in formal housing with basic services connected to their homes (e.g. water, flushing toilets, and electricity). However, job security in these towns is directly linked to the price of natural resources (iron ore in the case of Kathu), which can force mining companies to take austerity measures in order to remain profitable. This suggests that resource-dependent towns in frontier regions are economically at the mercy of global influences and fluctuations, which can lead to the resource curse, as it is referred to in the literature.

In most towns, economic development has actually declined, with more people earning low incomes. However, corporate social responsibility programs (social labour plans and government working with these companies) have improved basic household infrastructure and basic services connected to these houses. Nonetheless, towns like Kathu, which are in the boom phase of the mining cycle, show an upward curve in terms of household and individual income. The improvements in all of these towns were noticeably measurable over the ten-year measurement period. Lastly, it is important to prioritize the enhancement of skills for the youth in small mining towns in the Northern Cape, as education and skills levels are relatively low and improving them would enable youth to be more economically active.

In the study, each dimension had indicators representing that dimension. Towns were ranked according to their performance in each indicator, from best to worst. The indicators within the same dimension were grouped together and their average rankings were calculated to rank each town according to the dimension tally. Thus, the ranking is a composite indicator index. This index shows that the level of socio-economic wellbeing is directly linked to the stage of the mining cycle in which the town finds itself: the level of wellbeing is higher where the mining operation is still viable and profitable. Towns that experienced the boom-bust cycle were near the bottom of the ranking, except for Kleinzee and Koingnaas. In the case of Kleinzee and Koingnaas, this is due to the number of people who migrated away from the area and the fact that the remaining small population is highly developed (housing and economically), and therefore can afford to live there without being dependent on the mining sector any longer. The level of socio-economic well-being seems to fluctuate with the life cycle of the mine, and no cases were found where a town flourished after mine downsizing or closure. With the towns in the Northern Cape being so remote, expenditure on tourism seems futile, and investment in agriculture is risky as alternative forms of development because of the current drought, which compounds the issue of water scarcity and quality. Future research should focus on the general CSR policies, as well as the specific Social and Labour Plans (SLPs) developed by mining companies operating in the study area since the promulgation of the Mineral and Petroleum Resources Development Act in 2002.

References

Atkinson, G., & Hamilton, K. (2003). Savings, growth and the resource curse hypothesis. *World Development, 31*(11), 1793–1807.

Black, D., McKinnish, T., & Sanders, S. (2005). The economic impact of the coal boom and bust. *The Economic Journal, 115*, 449–476.

Bravo-Ortega, C., & De Gregorio, J. (2007). The relative richness of the poor? Natural resources, human capital and economic growth. In D. Lederman & W. Maloney (Eds.), *Natural resources, neither curse nor destiny* (pp. 71–99). World Bank Publications.

Brown, R. B., Dorius, S. F., & Krannich, R. S. (2005). The boom-bust-recovery cycle: Dynamics of change in community satisfaction and social integration in Delta, Utah. *Rural Sociology, 70*(1), 28–49.

Burger, P., Marais, L., & Van Rooyen, D. (Eds.). (2018). *Mining and community in South Africa.* Routledge.

Dashwood, H. S. (2012). *The rise of global corporate social responsibility: Mining and the spread of global norms.* Cambridge University Press.

Freudenburg, W. R. (1992). Addictive economies: extractive industries and vulnerable localities in a changing world economy. *Rural Sociology, 57*(3), 305–332.

Freudenburg, W. R., & Wilson, L. J. (2002). Mining the data: Analysing the economic implications of mining for nonmetropolitan regions. *Sociological Inquiry, 42*(4), 549–575.

Hajkowicz, S. A., Heyenga, S., & Moffat, K. (2011). The relationship between mining and socio-economic well-being in Australia's regions. *Resources Policy, 36*(1), 30–38.

Halseth, G. (1999). "We came for the work": situating employment migrations in B.C.'s small, resource- based communities. *Canadian Geographer, 43*(3), 363–381.

Hamman, R., & Kapelus, P. (2004). Corporate social responsibility in mining in Southern Africa: fair accountability or just greenwash? *Development, 47*(3), 85–92.

Hanekom, F. (1976). *Die hulpbronfront: 'n gevallestudie van delfstowwe se invloed op streekontwikkeling in Noord- en Noordwes-Kaapland* [Doctoral dissertation]. University of South Africa (UNISA).

Ivkovic, A. F., Ham, M., & Mijoc, J. (2014). Measuring objective well-being and sustainable development management. *Journal of Knowledge Management, Economics and Information Technology, 4*(2), 1–29.

Kapelus, P. (2002). Mining, corporate social responsibility and the "community": the case of Rio Tinto, Richards Bay Minerals and the Mbonambi. *Journal of Business Ethics, 39*, 275–296.

Kleynhans, E. (2012). The effect of declining goldmine production levels on employment in the Dr. Kenneth Kaunda District Municipality. *Tydskrif vir Geesteswetenskappe, 52*(4), 678–703.

Lawrie, M., Tonts, M., & Plummer, P. (2011). Boomtowns, resource dependence and socio-economic well-being. *Australian Geographer, 42*(2), 139–164.

Littlewood, D. (2004). Cursed communities? Corporate social responsibility (CSR), company towns and the mining industry in Namibia. *Journal of Business Ethics, 120*, 39–63.

Marais, L. (2013). Resource policy and mine closure in South Africa: The case of the Free State Goldfields. *Resources Policy, 38*(3), 363–372.

Marais, L. (2018). Housing policy in mining towns: issues of race and risk in South Africa. *International Journal of Housing Policy, 18*(2), 335–345.

Marais, L. (2023). *The social impacts of mine closure in South Africa: housing policy and place attachment*. Routledge.

Marais, L., & Cloete, J. (2009). Mining and housing: the case of the "village under the trees" (Kathu – South Africa). *Town and Regional Planning, 55*, 32–38.

Marais, L., & De Lange, A. (2021). Anticipating and planning for mine closure in South Africa. *Futures, 125*, 102669.

Marais, L., McKenzie, F., Deacon, L., Nel, E., Van Rooyen, D., & Cloete, J. (2018). The changing nature of mining towns: Reflections from Australia, Canada and South Africa. *Land Use Policy, 76*, 779–788.

Marais, L., Campbell, M., Denoon-Stevens, S., & Van Rooyen, D. (Eds.). (2021). *Mining and community in the South African Platinum Belt: a decade after Marikana*. Nova Science Publishers.

Marais, L., Cloete, J., & Lenka, M. (2022a). The plight of mining cities in South Africa: planning for growth and closure. *Cities, 130*, 103965.

Marais, L., Ndaguba, M., Mmbadi, E., Cloete, J., & Lenka, M. (2022b). Mine closure, social disruption and crime in South Africa. *The Geographical Journal, 188*(3), 383–400.

Nardo, M., Saisana, M., Saltelli, A., & Tarantola, S. (2005). *Tools for composite indicators building*. European Commission, Directorate-General, Joint Research Centre.

Office of the Premier of the Northern Cape. (2012). *Northern Cape provincial development and resource management plan/Provincial Spatial Development Framework (PSDF)*. Department of Cooperative Governance: Human Settlements and Traditional Affairs.

Organisation for Economic Cooperation and Development (OECD). (2011). Compendium of OECD well-being indicators. .

Organisation for Economic Cooperation and Development (OECD). (2012). Measuring well-being and progress. .

Organisation for Economic Cooperation and Development (OECD). (2016). *Society at a glance 2016: OECD social indicators*. OECD Publishing.

Organisation for Economic Cooperation and Development (OECD). (2020). *"Executive summary", in How's Life? 2020: Measuring Well-being*. OECD Publishing.

Rogerson, C. M. (2012). Mining-dependent localities in South Africa: the state of partnerships small town local development. *Urban Forum, 23*(1), 107–132.

Saisana, M. (2004). *Composite indicators – a review*. Proceedings of the second workshop on Composite Indicators of Country Performance held 26–27 February 2004. OECD.

Sesele, K., & Marais, L. (2022). *Mine closure, women and crime in Matjhabeng, South Africa*. Geographical Research. https://onlinelibrary.wiley.com/doi/epdf/10.1111/1745-5871.12563

Shao, S., & Yang, L. (2014). Natural resource dependence, human capital accumulation, and economic growth: a combined explanation for the resource curse and the resource blessing. *Energy Policy, 74*, 632–664.

Smith, M. D., Krannich, R. S., & Hunter, L. M. (2001). Growth, decline, stability and disruption: a longitudinal analysis of social well-being in four western rural communities. *Rural Sociology, 66*(3), 425–450.

Spocter, M. (2021). Kimberley: The diamond city has lost its sparkle. In A. Lemon, R. Donaldson, & G. Visser (Eds.), *South African urban change three decades after apartheid: homes still apart?* (pp. 215–232). Springer.

Stedman, R. C., Parkins, J. R., & Beckley, T. M. (2004). Resource dependence and community well-being in rural Canada. *Rural Sociology, 69*(2), 213–234.

Stijns, J. C. (2006). Natural resource abundance and human capital accumulation. *World Development, 34*, 1060–1083.

Stilwell, M. (2011, November 18). Diamond mines are not forever. *Mail & Guardian.*

Tonts, M. (2010). Labour market dynamics in resource dependent regions: an examination of the Western Australian Goldfields. *Geographical Research, 48*(2), 148–165.

Tonts, M., Plummer, P., & Lawrie, M. (2011). Socio-economic wellbeing in Australian mining towns: a comparative analysis. *Journal of Rural Studies, 28*, 288–301.

Warhurst, A., & Noronha, L. (2000). *Environmental policy in mining: corporate strategy and planning for closure.* Lewis.

Chapter 8
Complex Town-Gown Relationships and the Paradoxes of Studentification in the South African Traditional Authority Lands

Nothile Ndimande

8.1 Introduction

Throughout history, universities have had to define their roles in society. Questions, ambiguities, and ambivalences arise from how universities view themselves and are perceived by others, but the university's communion with its surroundings remains the most crucial aspect. The town-gown relationship is unique to the institution's geographic location (Moore, 2014; O'Mara, 2012). These relationships cannot be easily classified as either positive or negative. Instead, they may fall across a spectrum that extends beyond the dichotomies of good and bad. They also depend on time, context, and the parties' efforts to find common ground. Over the years, there has been a proliferation of literature in student geography that examines studentification and town-gown relationships. For a long time, town-gown research has become more prevalent in the global north. These studies, particularly in the global north, have emphasised the relationships between the town and the gown, with a focus on the dynamics presented by the interactions of various actors, including municipal authorities, property developers, residents, businesses, university officials, and students. There is also a growing interest both in major cities and towns in the global south (Mabin, 2018; Hendricks & Flaherty, 2018; Thakrar, 2018). However, only a small number of studies examine the town-gown relationships in areas under the South African traditional authorities. Therefore, this case study aimed to examine the relationship between the University of Zululand (gown) and the KwaDlangezwa community (town). This chapter begins with an overview of the

N. Ndimande (✉)
Department of Geography and Environmental Studies, University of Zululand, Richards Bay, South Africa

Department of Geography and Environmental Studies, Stellenbosch University, Stellenbosch, South Africa
e-mail: ndimanden@unizulu.ac.za

© The Author(s), under exclusive license to Springer Nature Switzerland AG 2023

R. Donaldson (ed.), *Socio-Spatial Small Town Dynamics in South Africa*, GeoJournal Library, https://doi.org/10.1007/978-3-031-37142-4_8

study's background, followed by a discussion of the existing literature and an outline of the research methodologies employed. The chapter then presents the results and discussion and concludes by summarising the main findings and recommendations.

8.2 Background: A University on Traditional Land

Universities in South Africa were founded on the core-periphery ideology, with the majority of historically privileged universities located in the core (major cities with thriving economic activity) and the historically disadvantaged institutions (HDIs) located in the periphery (small towns and villages). This geography has influenced the nature of the relationships between various universities and their surrounding communities. Homeland and ethnically based universities, such as Zululand, were instruments for achieving racial segregation in South Africa, with location-based responsibilities (Bank, 2018; Nkomo & Sehoole, 2007). KwaDlangezwa, an area under the traditional authority of the Mkhwanazi, became the location of the University of Zululand. When the University was founded, black South Africans viewed it as an apartheid project whose purpose was to perpetuate the ideologies of segregation. However, the then Zulu King, Bhekuzulu kaSolomon, persuaded people to accept the University as a great opportunity for Africans (University of Zululand, 1999). In January 1960, the University established an advisory board, which was one of the committees of the University Council, which was composed of a representative of the Zulu Royal Household and a local traditional authority representative. The primary purpose of having a representative was to foster relationships between the University and the surrounding community. The purpose of this advisory body was to help the predominantly white university administration understand the local context in which the University operated (University of Zululand, 1968). The structure seems to have vanished in the 1970s, as it is absent from University records. Currently, the University Council, a governing body comprising representatives from local stakeholders, includes a representative of the Mkhwanazi Traditional Authority.

The University's expansion led to the development of Vulindlela Township in 1970, literally on the doorstep of the University. This became a dormitory for the University's African staff and later expanded to house employees of the KwaZulu government (a former homeland).[1] For a long time, the University of Zululand was the only university in the north-eastern part of KwaZulu-Natal province, and it continues to be the only residential university in this region. The University is located in KwaDlangezwa (depicted in Fig. 8.1) in the City of uMhlathuze (a local municipality), within the King Cetshwayo District. It is part of ward 30, which has an

[1] Interview with Prof Vika Gabela, a former professor in the Faculty of Education, who, during his tenure at the University of Zululand, served in many committees, including the University Council.

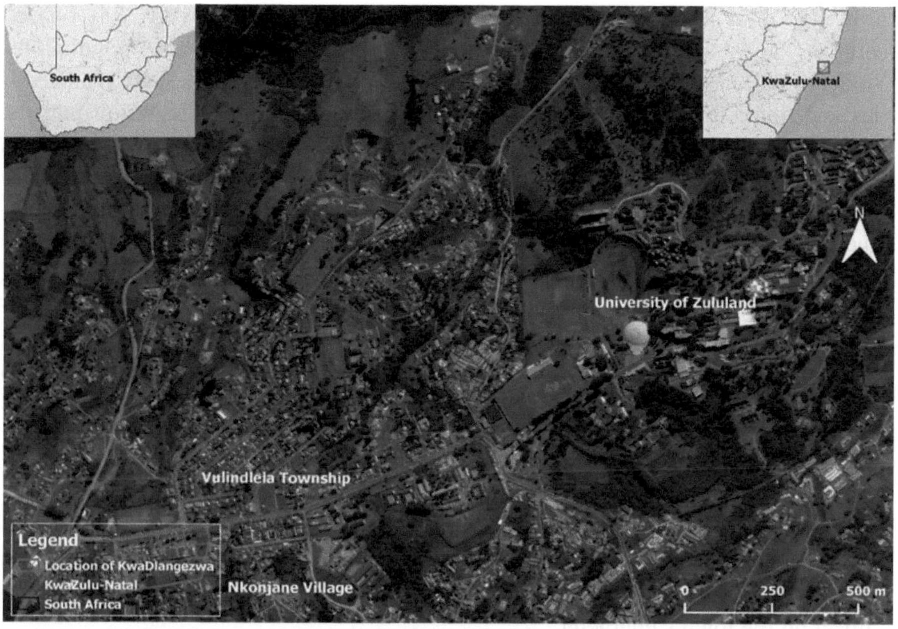

Fig. 8.1 KwaDlangezwa. (Source: Author, 2023)

estimated population of 11,355 (City of uMhlathuze, 2021). In 2021, the student population was approximately 17,000, with the majority of students residing in off-campus housing in KwaDlangezwa.

8.3 Related Literature: Studentification and Town-Gown Relationships

Town-gown connections are critical to the institution's success (Harasta, 2022). Universities develop a culture and give character to the college town; they thus become place makers and they significantly influence processes that produce space (Liu, 2017; Moore, 2014; Gumprecht, 2003). They also have an important role in the local economy as anchor institutions (Sladek, 2019). Socio-economic problems have always demonstrated that universities cannot separate themselves from people (Martin et al., 2005). As a result, questions have been raised about whether universities should play a role in tackling socioeconomic issues and if so, whether they have the resources to do so (O'Mara, 2012). The response depends on a number of factors, especially town and gown leadership, their strategic visions and missions, the town's socio-economic conditions and the university's resources.

The students' presence causes changes in the town (Smith, 2019). Studentification is instrumental in connecting the town and the gown. As university enrollment grows, so does the need for student housing. Universities frequently rely on off-campus housing to accommodate their expanding student populations. Studentification serves as a conceptual framework for stakeholders such as universities and municipalities to plan, monitor, and manage student integration into the community, as well as to foster beneficial town-gown relationships (Smith & Fox, 2019). In charting the evolution of this conceptual framework, Revington (2021) identifies four trajectories of studentification: classical, corporatised, de-studentification, and post-studentification. Students living in shared housing and the conversion of single-family homes to student housing are both examples of classical studentification (Gregory & Rogerson, 2019; Smith & Holt, 2007). The corporatised refers to developer-provided purpose-built student accommodation (PBSA), which often has high-end amenities and services to meet the demands of students (Reynolds, 2020; Kinton et al., 2018; Mulhearn & Franco, 2018; Hubbard, 2009). De-studentification refers to students vacating neighbourhoods that were once dominated by students (Kinton et al., 2016). In studentified neighbourhoods, post-studentification refers to a blend of student and non-student inhabitants (Revington, 2021). In the global south, one may add slummification, where there is an extensive expansion of unplanned and unregulated shack-like structures to accommodate students (Ndimande, 2018). These paths show that studentification has a significant impact on the town's character.

While studentification is crucial for connecting the town and the gown, it is also a source of a variety of concerns, such as increased traffic, loud noise, alcohol, and late-night parties (Vandegrift et al., 2012; Miller & Rivera, 2005). For example, the annual social events of the students in Fairfield, Connecticut, which the citizens regarded as a nuisance, sparked a town-gown conflict (Aggestam & Keenan, 2007). Stakeholder engagement sessions revealed additional dimensions of locals' dissatisfaction, such as the leasing of beach properties to students. It was through these engagements that the municipality enacted ordinances on the lease of beach properties and the hosting of social events (Aggestam & Keenan, 2007).

Town-gown interactions, or lack thereof, can be a source of contention. Some universities have been accused of being isolationist or having an ivory tower ethos. They have isolated themselves by designing and constructing physical barriers to the outside world, such as fences and gates, or by erecting reclusive campuses and providing student housing and services on campus (Bunnell & Lawson, 2006; Martin et al., 2005). Isolationist attitudes birthed an antagonistic town-gown relationship, with universities seen as working hard to reinforce disunion (Mosier, 2015). Unfortunately, the absence of relationships prevents both universities and communities from addressing issues that affect them.

Numerous theories and concepts have been employed to explain the relationship between the town and the gown. Among these are the urban disorder theory and broken windows theory (Woldoff & Weiss, 2018), the marriage metaphor (Gavazzi, 2015b; Gavazzi & Fox, 2015; Gavazzi et al., 2014), UniverCities (Baldwin, 2021) and the right to the city (Bank & Paterson, 2018). The theory of urban disorder examines the intrusion of students into cities and towns and the disorder resulting from their presence, such as vandalism, protests, and public drunkenness (Woldoff

& Weiss, 2018). Residents can respond either actively or passively to disturbances caused by students. Active response means that residents seek corrective intervention whenever students cause disorder, whereas a passive response indicates that residents accept disorder out of fear of aggravating those whose livelihoods depend on the university and the students. Intrinsically connected to the urban disorder theory is the broken windows theory. It seeks to dispel the myth that vandalism stems from cultural pathology by contending that crowd mentality, social inequalities, and anonymity can trigger inappropriate behaviour (Ansfield, 2020). In their interpretation of the broken windows theory Wilson and Kelling (1982) contend that there is a correlation between crime and disorder whereby one act of vandalism leads to more acts of vandalism, meaning that if a broken window is left unrepaired, it is likely that additional windows will be broken. Broken windows are frequently associated with a pronounced pattern of urban decay in the city that is characterised by transient populations (Harasta, 2022; Woldoff & Weiss, 2018). Rowan (2017) refutes the link between crime and disorder as an oversimplification and misinterpretation of the broken windows theory and contends that disorder increases a community's susceptibility to crime due to decreasing social controls or the ability of the residents to deal with nuisance and disobedience. This theory, Rowan adds, is not founded on empirical evidence but rather on the normative ideal of assigning police to maintain law and order. Most college towns feature what Gumprecht (2003) terms a student ghetto, a neighbourhood with decrepit housing populated primarily by students. Incivilities such as noise and public drunkenness are more prevalent in such areas as a result, and police have been asked to fix broken windows (bring order), while other residents have come to understand that disorder is ingrained in the college town's genetic structure (Woldoff & Weiss, 2018). In addition to the aforementioned misconstrued readings of the theory, Ansfield (2020) contends that the erroneous interpretation of the theory has been used to legitimise racialised law enforcement.

The marriage metaphor suggests that town-gown partnerships are analogous to arranged marriages in which the partners remain married regardless of their circumstances (Gavazzi, 2015a; Gavazzi & Fox, 2015). University presidents or vice-chancellors play critical roles in establishing and maintaining town-gown partnerships to guarantee that universities have a good impact on the town's many sectors, especially the economy (Weill, 2009). To explain distinct types of town-gown relationships, Gavazzi et al. (2014) employed typologies adopted from marriage scholarship and used examples of universities in the United States and Canada to demonstrate these relationships. Effort and comfort are integral components of all types of relationships. First, there is the devitalised relationship, in which the community and university previously had a good relationship, but it has deteriorated because of one party's disappointment. Devitalisation can be seen in the failed talks between Mansfield and Ohio State University, in which Mansfield tried to acquire the regional campus of Ohio State University located in Mansfield (Gavazzi et al., 2014). The second type of relationship is the conflict relationship, which, despite having certain good qualities, is unstable and marked by disagreements and confrontations. This is a tumultuous town-gown relationship in which one party's conduct may spark conflicts and confrontations with the other. The ongoing conflict between Kingston and Queens University in Ontario was precipitated by the provincial government's decision (in the 1950s) to grant the university the right to expropriate property to meet the expanding demand for student housing (Gavazzi et al., 2014). Third is the traditional relationship, which is where the university and the community function in silos with little interaction. This relationship is put to the test whenever there is development, such as the expansion of university residences. An example of this is Mount Allison University and Sackville in New Brunswick, who remain distant despite their small populations (Gavazzi et al., 2014). Finally, there is a harmonious partnership. This is a desirable situation in which the university and community work together to enhance their relationship, and there is extensive sharing of activities and resources, resulting in an ideal town-gown partnership (Gavazzi & Fox, 2015). Clemson University and the city of Clemson are considered to have a harmonious relationship due to their shared visions, well-established communication channels, and collaborative activities (Gavazzi et al., 2014). It is essential to recognise that variations in these relationships are a function of time, context, and size of both the town and gown. Variation in time implies that relationships are unpredictable and may fluctuate between favourable and unfavourable states. The variations in context and size are largely influenced by the disparities between town and gown circumstances, which may favour one over the other because their playing fields are unequal. It is imperative to explore such relationships in the global south. In South Korea, for example, town-gown relationships have a different dynamic. The power lies with the residents, specifically the landlords of student housing (Jang, 2020). These landlords' source of income is student housing; consequently, they oppose anything that threatens their livelihood. They have the power to approve or reject a university's expansion request. Other examples worthy of notice are town-gown relationships in post-conflict African countries. In Kenya and Côte d'Ivoire, town-gown experiences have been investigated using post-elective conflict as a central theme (Johnson, 2017). Following their disputed national elections, violence erupted in both countries, and universities were attacked, either because communities and universities had strained relationships due to perceived neglect, or because universities were not seen as valuable institutions. Universities typically wait for a crisis to occur before responding. Their reactions are sometimes a little late (Harasta, 2022). At the end of the conflict, Kenya worked on a model to improve community relations, whereas Côte d'Ivoire built walls and fences to protect itself (Johnson, 2017). These two African cases present conflict-oriented relationships where disconnections seem to be the main source of conflict.

UniverCities is a term coined by Baldwin (2021) to describe the labyrinthine influence of universities on cities and towns. The primary concern is that universities could consume and influence the growth and functionality of cities. They are taking over urban areas and have become parasites, suffocating the residents and small businesses, necessitating a review of these town-gown relationships. The tax exemption of universities has always been a source of contention between town and gown in the United States (Weill, 2009; Gallo & Davies, 2008). The classification of universities as non-profit institutions when they play a significant economic role in sectors such as the real estate market creates a paradox. The UniversCities concept raises questions regarding urban politics which among others includes unjust displacement of the low-income residents. For example, Hispanic and African American communities felt that they were unfairly pushed out of the city because of the expansion of the University of Southern California in Los Angeles. This situation resulted in community mobilisation against university-based social injustices (Dreier, 2013). Another example was Trinity College, Hartford in Connecticut, which constructed fences on a public street and regulated access for campus safety (Baldwin, 2021), or the University of South Carolina's hunger for the slum clearance project (Kahler & Harrison, 2020). Another concern is campus policing, in which the authority of university security operations is extended beyond the territory they are tasked with protecting, and in which the policing of black people becomes a political issue (Baldwin, 2021). These instances demonstrate a lack of trust and a power disparity, with the gown appearing to have more authority than the town.

Henri Lefebvre's concept of right to the city was adopted by (Bank & Paterson, 2018) to analyse student protests in East London, the second city in the Eastern Cape, South Africa, where students blocked roads to protest living conditions. Students from the city campuses of Fort Hare,[2] Walter Sisulu[3] and South Africa[4] complained about unscrupulous landlords who charge exorbitant rents and have no desire to improve living conditions, while city residents expressed displeasure at the presence of students in their neighbourhoods, often viewing them as deviants. Concerns centred on the inadequate responses from university authorities and the failure of the municipality to facilitate dialogue between the key stakeholders, including students, university administrators, landlords, and other concerned parties. There was also a grave concern that city officials had missed an opportunity to implement place-based urban reform and address the emerging needs of the city's youth (Bank & Paterson, 2018).

South African universities are no different than their international counterparts. Some have isolated themselves from their surroundings, citing student safety as justification (Mabin, 2018; Hendricks & Flaherty, 2018). Others have reached out to their host towns to build stronger relationships. Two cases in South Africa of city-based universities demonstrate their involvement in city revitalisation by developing urban precinct projects. The University of the Witwatersrand (Wits), one of South Africa's historically most advantaged universities, has had its own history of connections, stumbles, and disconnections with the city. Mabin (2018) explores the relationship between Wits and Johannesburg. The presence of students and their use of city social facilities has always connected the University to its immediate surroundings, but safety concerns were raised in the 1990s, prompting the university to

[2] The University of Fort Hare's main campus is in Alice, a small rural town in the Eastern Cape.

[3] Walter Sisulu University's main campus is in Umtata, a town in the Eastern Cape.

[4] The University of South Africa is South Africa's biggest distance learning institution with branches in various towns and cities across the country.

build walls and install biometric verification infrastructure to control access and isolate the university from its surroundings. There have been some beneficial partnerships between the town and gown that resulted in the development of the Hillbrow Health Precinct, which revitalised the city's dilapidated areas (Mabin, 2018). The University of Pretoria expressed an interest in forming collaborations with relevant stakeholders to promote accessibility and the development of an outward-focused campus, citing the University of Pennsylvania as a model (Hendricks & Flaherty, 2018). As an anchor institution, the University explored the feasibility of developing a Hatfield Campus Precinct, and there were numerous positive attributes that made the project achievable, such as land ownership in the Hatfield area, visionary leadership, and engagement with local stakeholders. Both of these universities have made deliberate investments in urban revitalisation projects that benefit both the towns and gowns. The same cannot be said for the main campus of the University of Fort Hare, a previously disadvantaged South African university. Apartheid laws deliberately produced a dichotomic relationship between Alice (town) and the University of Fort Hare, a rural-based university. In February 1916, missionaries founded the institution for African students (Matthews, 1957). Initially, the University had a modestly diverse student body, with predominantly black, a few white, and African students from outside South Africa. In the late 1950s, the apartheid administration redefined the university as an ethnic institution for South Africans who spoke isiXhosa. This move was met with resistance, which saw political activism ensue within the university. Political activism at the University upset Alice's residents, who, at the time, were predominantly white (Thakrar, 2018). The effects of apartheid continue to persist even in the post-apartheid era. Literature and government development initiatives reveal ingrained disconnections, notably their reluctance to consider Alice a university town (Ntema, 2021). The disconnections are also indicated by the absence of road signage which recognises the University's presence within the town and the University's controlled access system (Thakrar, 2018). However, the University and the Raymond Mhlaba Local Municipality have subsequently acknowledged their estrangement and stated their desires to rectify the disconnections through the University's strategic plans and the Municipality's spatial development plans, such as the Alice Regeneration Project, which aims to encourage development that will benefit both the town and the university (Ntema, 2021; Thakrar, 2018). However, the town's unwillingness to recognise the University's existence through simple things such as signage, and the University's enclaved and gated design continues to separate the town from the university (Thakrar, 2018). It thwarts efforts to bridge the gap between the two. Fort Hare and Rhodes University are located in the same province and had intersected at some point in their timelines, but they have had dissimilar town-gown experiences. Rhodes University, a historically privileged rural university, was founded in 1904 to advance a colonial agenda in South Africa. A proposal was pitched to the Grahamstown City Council to establish a university resembling Oxford University (Maylam, 2016). Since its inception, the university has been instrumental in fostering English influence in Grahamstown and beyond (Maylam, 2005). The town and gown histories were interwoven with the history of apartheid-era South Africa. The

University was regarded as an accomplice to apartheid and segregation measures (Hendricks & Vale, 2005). In post-apartheid South Africa, Grahamstown has seen studentification and racial integration. Yet, the growth of the student housing market has made access to the housing market problematic for some residents (Irvine, 2021). Notably, in 2014–15, the University and Grahamstown became sites of activism where marginalised groups from the community (the Unemployed Peoples Movement) and students (mainly blacks) united in their desperation to confront post-apartheid unemployment, a lack of service delivery and the remnants of colonialism (O'Halloran, 2018).

8.4 Research Methodology

To investigate the town-gown relationship between the University of Zululand and the KwaDlangezwa community, the research employs interpretivism, which at its core embraces a relativist ontology and subjective epistemology, allowing the researcher to view the world from the perspective of the research participants (Rashid et al., 2019). Qualitative research recognises the complexities of human thought and behaviour, and the researcher recognises that thoughts and behaviours are the result of a number of factors that contribute to life experiences (Roller & Lavrakas, 2015). Several town-gown studies have employed qualitative research methods, including ethnographic case study (Aggestam & Keenan, 2007), the phenomenological approach (Harasta, 2022), and the case study approach (Advocat et al., 2016). This research employed a case study approach because it recognises the uniqueness of each case and captures the context, dynamics, and practices at play (Aailtio & Pia, 2010; Stake, 1995). Case studies examine one or a small number of instances of a phenomenon to shed light on relationships, processes, events, or experiences (Denscombe, 2010). Community and University officials were asked to describe their perceptions and share their experiences of their respective relationships. To select research participants, a purposeful sampling method was used. This is a method by which the researcher, based on prior knowledge, can consciously select a sample to ensure that the entire range of items or individuals is represented (Denscombe, 2010). Thus, based on the researcher's knowledge of the University and the KwaDlangezwa community, a purposive sample was selected to ensure that a diverse group of participants was included in the study. Participants from the community were chosen on the basis of their interactions with the University. Interviews, both unstructured and semi-structured, were used to collect data. This flexibility enables the researcher to move back and forth between semi-structured and unstructured interviews, and allows participants to express themselves freely (Denscombe, 2010). This follows the practice in town-gown studies, such as Woldoff and Weiss (2018), where qualitative interviews were used to assess residents' perceptions of how studentification affected their neighbourhoods.

This study recruited 21 research participants: 1 serving university official, 4 retired university officials (three of whom are KwaDlangezwa residents), 2 community leaders (one is currently serving as a leader, while the other's tenure has recently concluded), and 15 residents of KwaDlangezwa. All four retired university officials were heads of various sections of the University. The majority of the residents interviewed were either street vendors or landlords. The interviews were conducted at the participant's residence, or at a location of their choosing, or at the makeshift market stalls located at the University's main entrance. In addition to interviews, various documents including university calendars, magazines, policy documents, and strategic plans were utilised to gather information on the University's past, particularly regarding land ownership and community relations. Print and online newspaper articles were utilised to identify events, such as protests, and to look for the voices and narratives that were reflected in these stories.

As Saldaña (2014) notes, the objective of data analysis is to reveal to others, through novel perspectives, what we have observed and learned about the human condition. The analysis of data begins with the organisation of raw data and audio recordings of all conducted interviews. In this study, all the audio recordings were transcribed, and the transcripts labelled, annotated, and filed into two groups: gown (university officials) and town (residents, landlords, street vendors, and other entrepreneurs). The transcribed materials were supplemented with newspaper clippings, printed online news articles, and other printed materials. Interview questions were used to create anchor codes, and these were recorded on the MS Word table. To identify patterns and similarities, the following codes were borrowed from Saldaña (2014): attributes, emotions, and versus. The following section presents the findings from examining the connection between town and gown. Three broad themes encapsulate these study results: land, studentification, and relationships.

8.5 Land Dispossessions or Donations, and the Ownership Dilemma

Higher education institutions in South Africa were already segregated on the basis of language (English and Afrikaans) and race (Black and White) before apartheid became the country's official policy in 1948 (Davies, 1996). The University of Zululand was established after the Extension of University Education Act 45 of 1959 was passed. This act made provision for the establishment of institutions based on ethnicity, which the government hoped would train black leaders who would care for their respective homelands (Davies, 1996; Christopher, 1994). There is a strong connection between the University of Zululand's history and land-based laws such as the Native Land Act of 1913 and the Native Trust and Land Act of 1936. The 1913 Native Land Act restricted African land ownership to scheduled African reserves and permitted Africans to own 7% of the land (Feinberg, 1993). The Act separated rural lands into black and white zones. The Native Land Act of 1936 was an extension of the Land Act of 1913. It sought to acquire more land for Africans. The Native Trust Act established the South African Native Trust, later renamed the South African Development Trust, to administer and manage lands designated for black people (reserves). The act authorised the Trust to expropriate land from black people when necessary (Schoch et al., 1938). It also transferred land-related authority from traditional leaders to the Governor-General. In order to make room for the University of Zululand, the South African Development Trust expropriated land occupied by the KwaDlangezwa communities. At that time, the Mkhwanazi Traditional Authority had no authority over the so-called trust lands in KwaDlangezwa. The traditional authorities and the residents had no choice but to locate alternative sites for those who were moved from the site designated for university construction. They were also compelled to excavate the graves of their deceased loved ones, and this became a source of embitterment as their lives were disrupted.[5] The University was granted permission to occupy (PTO) the land in 1977.[6] This was evidence that the University had been legally allotted land. The institution was initially viewed with resentment and hostility because people were abandoned after their removal from the KwaDlangezwa site. It was also looked at with suspicion because it was regarded as an imposition on the Mkhwanazi area.[7]

When discussing the expropriation of land at the time of the University's founding, some of the locals became extremely emotional. One of the residents of KwaDlangezwa recalled the removal of residents to make way for the University:

Community leader 1

> Let me tell you something about the land that was given to the University. It wasn't barren. It had homesteads, and people had fields and livestock. That's where our homestead was… We gave up our land for the University. When people were relocated, there was a saying that the University was a place where people would derive their livelihood… '*ikhona la eniyocosha khona uqadolo wenu'*, which literally translates to 'a place where you'll harvest indigenous or wild spinach'. A promise for a better life.

It is worth noting that the University claimed in its Self-Evaluation Portfolio (2010) that the land on which the University is built was a donation.

> The land on which the University is built was donated by the local community. Local residents at the time complained that they were displaced: *Kodwa nisisaphi?* (*where are you taking us to*?[8]) The response was 'ANYWHERE' (University of Zululand, 2010)

I argue that the word 'donation' was used to depoliticise and de-escalate the seriousness of the situation and the hardships that people endured during the 1960s removals. It also vaguely captured the community's devastation when they were left with no choice but to move.

In describing the University's relationship with the KwaDlangezwa community, Boughey (2014) contends that it is more a matter of fate than of choice. Communities had neither a voice nor a choice. They were compelled to accept the loss of land, and they had no ability to contest their removal.

---

[5] Prof Vika Gabela (ibid).

[6] Langa Nqengelele, a spokesperson of the former Minister of Higher Education in South Africa, cited by the City Press on the 29th October 2018.

[7] Prof Vika Gabela (ibid).

[8] Author's translation

In 1981, the government declared its intention to transfer the title deed to the University.[9] The apartheid government's decision to grant the University a title deed for land that the Mkhwanazi community viewed as ill-gotten sparked hostility. The traditional authorities confronted the University and cautioned it against pursuing the title deed without first consulting them. The University took immediate action to quell the brewing storm. Among the agreements and compromises that the University and the Mkhwanazi Traditional Authority reached were the following: firstly, the University offered to provide clean drinking water to households bordering its fence; secondly, the University consented to the opening of multiple entrances to facilitate community access to the University and the bus station. Due to the university's initial fencing, some locals had to walk considerable distances to access buses and taxis for their daily commute. During negotiations in the 1980s, five entrances were added to the University. Lastly, the University offered to construct a home for the then-*inkosi* Mkhwanazi. All of this was done to improve the University's relationship with the community.[10]

I should mention that, over time, communities have let their livestock graze and roam about campus. During the day-to-day flurry of academic activities, it is not unusual to see cows and goats grazing or wandering around campus (as shown in Figs. 8.2 and 8.3). In 2017, the University issued a notice to seize the local animals and impose a fine of R100 ($5 USD) per day (Wicks, 2017). However, the statement was withdrawn by the University and livestock continues to graze freely within the University premises.

At the end of the 1980s, the University requested a strip of land from the Mkhwanazi Traditional Authority in order to construct a new gate that would allow security staff to monitor incoming and outgoing traffic and enhance campus safety. The post office, a public facility located immediately outside the university gate, was to be included in the University grounds as a result of the land grant. A village elder recounts the University's expansion and the acquisition of previously unoccupied land:

> The post office was located outside of the University's campus. The University gradually took over some community-owned land. We recently learned that the University intended to close the post office. They issued an ultimatum to the staff. I sometimes get the impression that the University invites trouble. People in KwaDlangezwa get their social grants from the post office; why would you close or relocate it if you care about them? These people don't know the history of this University.

In 1991, the government passed the Abolition of Racially Based Land Measures Act with repealed the Land Act and the Native Trust Act and in 1992 the South African Development Trust was also abolished (Weideman, 2004). The land was retained by the state after the repeal of the land-based laws and the South African Development Trust, and later the Ministry of Regional and Land Affairs transferred more than a million hectares of land to the homeland governments. Half a million hectares of that land was given to the Ingonyama Trust, an independent body established in the

[9] Prof Gabela (ibid).

[10] Prof Gabela (ibid).

Fig. 8.2 Livestock wandering around campus. (Source: Author, 2022)

former KwaZulu homeland to manage land on behalf of the Zulu King (Lynd, 2021). The University celebrated its 40th anniversary in 1999, and it reflected on Prof. Abram Nkabinde's[11] tenure (1977–1993). The University indicated that one of its key achievements was the acquisition of a title deed to 153 ha of university land (University of Zululand, 1999). Even though the University magazine proclaims the University's possession of the title deed, the University continues its drive to obtain the title deed.

The title deed problem came to the forefront because of the 2019 student protest over unsafe off-campus living circumstances and the necessity to construct additional residences for students. In a statement issued by the University Council on March 12, 2020, the chairperson expressed disappointment with the land issue and how the Traditional Authority, in their opinion, was delaying the process of obtaining the title deed.

[11] Prof Abram Charles Nkabinde was the first black vice chancellor and the longest serving leader of the University of Zululand.

Fig. 8.3 Livestock wandering around campus. (Source: Author, 2022)

> Engagements with *inkosi* are moving at a snail's pace thus delaying the whole project. It is a plea of the University Council that *inkosi* prioritises this matter because engagements were at an advanced stage for the University to get a title deed. It is therefore this engagement with the *inkosi* Mkhwanazi that is delaying the process of building the student residence (University of Zululand, 2020)

The University is awaiting the signature of *inkosi* Mkhwanazi in order to release the land for the construction of additional student residences because it cannot build on land it does not own (Macupe, 2019). In 2018, the University bestowed a second honorary doctorate on the late Zulu King, Zwelithini *Ka* Bhekuzulu, in recognition of his humanitarian efforts. This was received with scepticism by some community members. They viewed the honorary doctorate as a quid pro quo to acquire land. A national newspaper, City Press, published an online article titled 'It's Dr King as varsity needs land,' in which they questioned the University's intentions of honouring the King for the second time. The paper also indicated that the University had written a letter to the Mkhwanazi Traditional Authority requesting *inkosi* Mkhwanazi to support their request to transfer land from the Department of Land Affairs to the University. However for such a transfer to occur, the university needs support from local stakeholders including the Mkhwanazi Traditional Authority (Fengu, 2018).

During the interviews with the community leaders, it emerged that the area directly outside the University's main entrance is a source of contention. Some members of the traditional household and the taxi association dispute the area's ownership. This contestation highlights the ambiguities that exist in communal lands. Briefly, land adjacent to the University is attractive to individuals looking to do business in KwaDlangezwa. The parked minibus taxis, improvised shacks, shipping containers, and gazebos set up by vendors dominate the area around the University gate. The area is also a site of *inkosi* Phalane's monument. To commemorate *inkosi* Phalane, one of their forebears who served as King Cetshwayo's top commander at the battle of Isandlwana in 1879, the Mkhwanazi Traditional Authority built a monument in 2019. One of the local elders explained why the monument was placed close to the University's main gate.

Resident 3

> The memorial honouring *inkosi* Phalane was placed close to the University's main entrance since *inkosi* Phalane existed before the University of Zululand. Before the institution was founded, people used to reside in this area. One of the *inkosi* Phalane's homesteads was located on the current site of the University.

The presence of the community on campus and the University's presence in the KwaDlangezwa area are both evident at the University's entrance.

8.6 Studentification, Informality and Livelihood

The strongest connection between the University of Zululand and the locals has been studentification. The area around the University has evolved over time into a haven for informality. Between the mid-1990s and the mid-2000s, the University's student enrolment increased by almost 70%. The University's ability to provide on-campus housing has been curtailed due to this rapid surge. During the interviews, landlords revealed that in the late 1990s and early 2000s, a number of residents began to construct backyard structures on their homesteads to house stranded students. Midway through the 2000s, residents began constructing freestanding structures which, in turn, encouraged outsiders (from neighbouring towns and townships) to also come and build more structures. Due to a lack of regulation, oversight, and disconnections between the institution and its surroundings, the proliferation of these informal structures has produced an environment that resembles a slum (Ndimande, 2018). Given that most of these buildings are subpar, this disconnect has given rise to unhealthy living conditions for students.

Student housing provides the locals with a livelihood. Most of them have sold their plots of land—once used to cultivate sugar cane—to outsiders, including some University academics and employees (Ndimande, 2018). Local landowners recognised the community's reliance on students for survival. The reciprocity between landlords and tenants was valued by a graduate whose family-owned student housing.

Landlord 1

> Without students, there will be no student accommodation, and there will be no livelihood. The University is crucial to our financial security. The University needs students, I need students, and students need accommodation. We therefore depend on one another. Because of students, we have been able to expand and make investments. During the 2020 lockdown, things were difficult. I made no profit. Students left for home. I had to pay for things like electricity and security out of my own cash. It really was hard.

A landlord with backyard student accommodation explained why some students choose lodging in homesteads.

Landlord 2

> Students frequently approach us. They seek safety. Even when the place is full, they beg us to give them a place to stay. Remember, ours isn't anything spectacular. All they want is safety. Our students have always been well-mannered and have demonstrated a strong sense of respect for one another. They are like family.

In addition to the owners of student housing, there are numerous street vendors who cater to students. For more than 20 years, street vendors have been operating within the boundaries of the institution. In the late 1990s, University dedicated a spot where most of them could find shelter and storage. At the time, there was a decent working relationship between the vendors and the institution. During the interviews, street vendors indicated that those who arrived in the early 2000s had storage space and market stalls. More vendors were permitted to sell on campus, even though some didn't have market stalls. When the University reopened in 2020 following the COVID-related lockdowns, however, things changed. Vendors reported that the University forbade them from coming back to campus and relocated them to the front entrance.

Street Vendor 1

> There were no talks. We were just called (phoned) to collect our stuff and were told that the university wants to build new structures, and when we got here, the site was already fenced. I think it was in November-December. We were hoping that construction will probably take a year so that we can get back to campus, but now we see that nothing is going on. Even though the university is back to normal after covid, we are not back on campus. It ended there. There were no talks between us and the university. We thought that it was a temporary arrangement.

The fact that the University forbade them from selling on campus has incensed a good number of them. They feel that they were more successful financially on campus than they are outside the University's main gate.

8.7 Relationships: Dysfunctionality and Remonstration

Initially, the community didn't patronise the University.[12] This could be attributed chiefly to two factors. Firstly, there was no high school in the area and many youngsters did not study further than junior secondary school, and thus did not qualify to

[12] Prof Gabela (ibid).

study at the University. Secondly, during the construction of the University, the government had employed non-local artisans such as bricklayers, plumbers and electricians from provinces such as Gauteng and Limpopo which were part of the former Transvaal. The University became accepted when employment opportunities became available for the local community. As time went by people realised that there were opportunities for cleaners and security guards and kitchen and dining staff. The anger that people had towards the University began to subside.[13]

Boughey (2014) points out the debates on what constitutes the community within the context of the University of Zululand. Views differ, with some arguing that it refers to the Mkhwanazi people whose land was taken to build the University, while others argue that the University community is drawn from within a 50 km radius. During the interviews, past and present university officials characterised the relationship between the KwaDlangezwa community and the University of Zululand mostly as a misprision, as they felt that the University had isolated itself from its surroundings. A former employee of the University who interacted with both the community and the Mkhwanazi Traditional Authority admitted that there was no proper relationship which existed during their tenure:

Former university official 2

> A tour was organised for a new group of staff members, a local tour, and one staff member, a white one, asked if they were safe to travel through a township, and she said they needed an armed guard. Those were her words. Those tours showed me just the level of not antipathy, as it was in her case, fear, or a total lack of comprehension of where we were located and what was on our doorstep…

Despite all the things being offered at one time or another, there has never been a truly symbiotic relationship with the Mkhwanazi community.[14]

Unfortunately, the complicated and antagonistic connection between the gown and the town in KwaDlangezwa has not improved, and the situation between the townies and the gownies remains combustible. The countless protests that occurred between 2018 and 2022 are proof of this. Off-campus protests by students concerning unsafe living conditions and criminal activity in KwaDlangezwa have lately been witnessed by the University and the local community. 2019 saw students obstructing traffic and blocking the entrance to KwaDlangezwa (Govender, 2019). They then disrupted academic activities at nearby schools. Students complained that they were not heard during the protests and urged that the University and the owners of the student housing have a dialogue to address the issue of crime in the area. Additionally, students set a neighbourhood-serving satellite police station on fire (Singh, 2019). They alleged that during emergencies, the police did not respond. Students protested once more in March 2020 to voice their displeasure with the University's handling of their national student funding programme amongst other things. The demonstration moved to the main street outside the University, and the

---

[13] Prof Gabela (ibid).

[14] Interview with Dr. John Boughey, a retired director. He was responsible for community engagement and academic development at the University of Zululand.

students set a police vehicle on fire in the process. Students' ire was directed at the University and the South African Police Services (SAPS) in both cases because they felt that these two institutions were non-responsive.

Residents of KwaDlangezwa have marched to the University's main entrance in 2018, 2020, and 2021. They have expressed their displeasure with what they consider the University's 'neglect'. They have accused the University of employing personnel such as administrative assistants, cleaners, and security guards from 'outside' KwaDlangezwa instead of the locals. They contend that, in keeping with the original pledge, locals should be given preference in such positions. Additionally, the University should purchase basic essentials from local businesses rather than elsewhere.

Community Leader 1

> Nothing is more upsetting than witnessing trucks from major cities bringing toilet paper to the campus. What makes the Durban suppliers so unique? Why won't they allow locals a chance to provide such necessities? What ever happened to the *uqadolo* (wild spinach) that was promised to us so long ago? ... Why can't the University ringfence opportunities for the community, like, say, security services? The University may go to the next ring if the community cannot offer these services, which may include searching for services across the entire municipal area before expanding to Durban or other provinces.

Resident and a community elder 2

> Our relationship with the University has shifted...The KwaDlangezwa people are no longer given priority. The situation between us and the University necessitates the intervention of men and women with the fortitude and strength to confront the University. This is our University. We must fix it.

The complex relationships and tensions between the town and the gown have been a liability for both parties. Protests are an indication of dysfunction, which, if left unattended, will stymie any attempt to build future relationships.

8.8 Discussion

This study investigated the town-gown relationship between the University of Zululand and the KwaDlangezwa community. The study's central thesis is that the relationship between the University and the KwaDlangezwa community has always been antagonistic, with varying degrees of tolerance and confrontation. It revealed complexes, triggers, conflicts, and conundrums through time.

Firstly, land is a contentious issue in South Africa as a whole, necessitating that it be handled with care. The history of land acquisition, which was influenced by both pre-apartheid and apartheid laws, contributed to the first layer of complexity. One can use Gavazzi and Fox (2015) marriage metaphor to argue that the relationship between the University and the KwaDlangezwa people was more of a forced marriage than an arranged one, with elements of bridal capture (land). This was an arranged marriage, as the community did not approve of the union. Instead, they were convinced to accept the status quo in exchange for a better life. They were initially cynical and dissatisfied with the apartheid government's imposition of a university in their region and the expropriation of their land. Using marriage typologies, I would like to argue that the relationship between the University and the KwaDlangezwa community has always been conflictual. Even though this relationship may, according to Gavazzi and Fox (2015) and Gavazzi et al. (2014) display some positive attributes, it remains a battleground for the parties involved. Land acquisition and the policies that enabled expropriation of land created ever-lasting problems for the parties involved. They have been oscillating between the 'good times' and 'bad times' trajectories characterised by a kaleidoscope of reciprocity and animosity. The University granted the community access to its premises by opening additional entrance points. It also granted informal traders' access and permission to conduct business on campus, although this is no longer permitted. Most significantly, KwaDlangezwa community provides housing for students and sells basic foodstuffs. Their livelihoods depend on the University's existence and the students' presence. In the 2020's the removal of campus vendors has strained the relationship between vendors and the University.

Noticeable are the triggers which detonate the community's hostile reactions towards the University, for example, the threat to confiscate livestock grazing on campus in 2017 and the order to the post office to vacate the University premises in 2021. The protests that occurred between 2018 and 2021 are indicative of the disharmony between the University and its surroundings, and this has been a huge liability which costs both the University and the community.

Studentification in KwaDlangezwa has been largely informal and has attracted informality into the university vicinity. This brings us to Woldoff and Weiss' (2018) theory of urban disorder, where disorder is evidenced in KwaDlangezwa by the proliferation of substandard student housing and informal traders' shacks at the University's main entrance. The community appears to be tolerant of the informality associated with studentification. This is mainly because studentification is their primary source of income. The passive responses referred to by Woldoff and Weiss (2018) will most probably ensure that informality continues to flourish. Interestingly, students have responded with protests, particularly in 2019 and 2022, demanding both the University and the police to protect them. Even though the town has repeatedly protested, they have never expressed discontent with the disorder caused by studentification. This is quite intriguing when one considers the student protests mentioned by Bank and Paterson (2018) whereby residents expressed resentment during student protests in East London.

The University has long expressed interest in acquiring the title deed for the University of Zululand. The enigma of the status, even the existence of the deed has muddled the issue of the ownership of the land on which the University of Zululand stands. When one considers the desperation to secure land rights in the Mkhwanazi area, numerous questions arise. The claim made in the UNIZULU'99 magazine that the University obtained a title deed during the tenure of Prof. AC Nkabinde raises doubts, among them the veracity of the claim.

Students have petitioned the University to extend its protective services beyond the University grounds due to the crimes committed against them. However, this has not occurred because the South African Police Service and the University have no agreement to collaborate beyond the University's boundaries. Resolution of this issue would hopefully aid in the protection of KwaDlangezwa's students. This argument makes sense in the context of South Africa. However, in other parts of the world, specifically the United States, expansion of the university's police department presents a complication associated with unfair policies towards black bodies in the cities, as Baldwin (2021) argues that this arrangement leads to the targeting and stereotyping of black students because they can be found in areas where white students predominate.

One hopes that when the University obtains the title deed, which is their objective, they will carefully negotiate with the community to avoid unwarranted tensions and conflicts. One further hopes that the University will not use the title deed to isolate itself from the community, but rather to build stronger relationships with the community.

8.9 Conclusion

There is an increased need to comprehend how history influences current town-gown relations. This deepening comprehension would aid in resolving conflicts between the University of Zululand and the KwaDlangezwa community. The intricate relationship between the two, described in this study, has focused on three key areas: land, studentification, and dysfunctionality and protests. The apartheid legislation, particularly land-based laws, influenced the character of the University's connection with its surroundings. It facilitated evictions and relocations, leading to animosity and hostilities. These tensions are mirrored in community protests against the University. The supply of substandard and unsafe off-campus accommodation has sparked violent student protests, which have periodically disrupted both the academic programme of the University and the lives of KwaDlangezwa residents. This study is important because it can be used as a reference point when a university and its local communities plan town-gown engagements. This research was limited to the KwaDlangezwa community, which is located close to the University. However, within the context of the University of Zululand, the town-gown research must be expanded to include other significant actors in order to cover the entire local municipality. The University's service area extends beyond KwaDlangezwa, so the scope of a study such as this should go beyond the local community and its traditional authority.

References

Aailtio, I., & Pia, H. (2010). Case study as a methodological approach. In A. Mills, G. Durepos, & E. Weibe (Eds.), *Encyclopedia of case study research*. Sage.

Advocat, J., Russell, G., & Mathews, M. (2016). Building links between town and gown: An innovative organisation in South-Eastern Melbourne. *Australian Journal of Primary Health, 22*(2), 71–76.

Aggestam, M., & Keenan, J. (2007). "Contraversations" constructing conflicts: Lessons from a town-gown controversy. *Business & Society, 46*(4), 429–456.

Ansfield, B. (2020). The broken windows of the Bronx: Putting the theory in its place. *American Quarterly, 72*(1), 103–127.

Baldwin, D. L. (2021). *In the shadow of the Ivory Tower: How universities are plundering our cities*. Boldtype Books.

Bank, L. (2018). Approaches to the university, place and development. In Bank, L., N. Cloete, & F. Van Schalkwayk (Eds.), Anchored in place: Rethinking the university and development in South Africa. Somerset West: .

Bank, L. & Paterson, M. (2018). The politics and pathology of place: Student protests, collective consumption and the right to the city in East London. In Bank, L., N. Cloete, & F. Van Schalkwayk (Eds.), Anchored in place: Rethinking higher education and development in South Africa. Somerset West: .

Boughey, J. D. (2014). *A comprehensive university at the heart of its communities: Establishing a framework for engagement*. University of the Free State.

Bunnell, G., & Lawson, C. T. (2006). A public university as city planner and developer: experience in the "capital of food planning". *Planning, Practice & Research, 21*(1), 25–43.

Christopher, A. J. (1994). *The Atlas of Apartheid*. Routledge.

City of Umhlathuze (2021). *Integrated development plan, (Idp) 2021/2022*. Richards Bay.

Davies, J. (1996). The state and the South African University system under apartheid. *Comparative Education, 32*(3), 319–332.

Denscombe, M. (2010). *The good research guide–for small-scale social*. Open University Press.

Dreier, P. (2013). Town versus Gown in Los Angeles. *New Labor Forum, Sage Publications, 22*, 90–93.

Feinberg, H. M. (1993). The 1913 natives land act in South Africa: Politics, race, and segregation in the early 20th century. *The International Journal of African Historical Studies, 26*(1), 65–109.

Fengu, M. (2018, April 29). *It's Dr King Zwelithini as varsity needs land*. City Press.

Gallo, R., & Davies, R. (2008). Research on the impact of Hbcus on African American communities. *Metropolitan Universities, 19*(3), 102–120.

Gavazzi, S. M. (2015a). Engaged institutions, responsiveness, and town-gown relationships: Why deep culture change must emphasize the gathering of community feedback. *Planning for Higher Education, 43*(4), 1–9.

Gavazzi, S. M. (2015b). For better and for worse: Understanding optimal campus-community relationships through the lens of marriage. *Metropolitan Universities, 26*(1), 147–154.

Gavazzi, S. M., & Fox, M. (2015). A tale of three cities: Piloting a measure of effort and comfort levels within town-gown relationships. *Innovative Higher Education, 40*(3), 189–199.

Gavazzi, S. M., Fox, M., & Martin, J. (2014). Understanding campus and community relationships through marriage and family metaphors: A town-gown typology. *Innovative Higher Education, 39*(5), 361–374.

Govender, S. (2019). Angry students block road in protest after robbers shoot sleeping man. *Times Live*.

Gregory, J. J., & Rogerson, J. M. (2019). Housing in multiple occupation and studentification in Johannesburg. *Bulletin of Geography Socio-economic Series, 46*(46), 85–102.

Gumprecht, B. (2003). The American college town. *Geographical Review, 93*(1), 51–80.

Harasta, J. (2022). Town & Gown-University and community leaders' perceptions on mutually beneficial relationships: An Urban American University case study. *Athens Journal of Mass Media and Communications, 8*(3), 161–178.

Hendricks, D. & Flaherty, J. (2018). Integrating the edges: University of Pretoria's neighbourhood anchor strategy. In L. Bank, N. Cloete, & F. Van Schalkwayk (Eds.), Anchored in place: Rethinking the university and the development in South Africa. Somerset West: .

Hendricks, F., & Vale, P. (2005). The critical tradition at Rhodes University: Retrospect and prospect. *African Sociological Review/Revue Africaine de Sociologie, 9*(1), 1–13.

Hubbard, P. (2009). Geographies of studentification and purpose-built student accommodation: leading separate lives? *Environment and Planning A, 41*(8), 1903–1923.

Irvine, P. M. (2021). From Grahamstown to Makhanda: Urban frontiers and challenges in a post-apartheid city. In *South African urban change three decades after apartheid: Homes still apart?* Springer.

Jang, A. (2020). University-community relations in urban regeneration. *Journal of Asian sociology, 49*(2), 163–192.

Johnson, A. T. (2017). Conflicted communities, contested campuses: A cross-case comparison of community engagement at two African Universities in conflict contexts. *The Review of Higher Education, 41*(1), 61–89.

Kahler, S., & Harrison, C. (2020). 'Wipe out the entire slum area': University-led Urban Renewal in Columbia, South Carolina, 1950–1985. *Journal of Historical Geography, 67*, 61–70.

Kinton, C., Smith, D. P., & Harrison, J. (2016). De-studentification: Emptying housing and neighbourhoods of student populations. *Environment and Planning A, 48*(8), 1617–1635.

Kinton, C., Smith, D. P., Harrison, J., & Culora, A. (2018). New frontiers of studentification: The commodification of student housing as a driver of urban change. *The Geographical Journal, 184*(3), 242–254.

Liu, C. (2017). *University spatial development and urban transformation in China*. Routledge.

Lynd, H. (2021). The peace deal: The formation of the Ingonyama Trust and the Ifp decision to join South Africa's 1994 elections. *South African Historical Journal, 73*(2), 318–360.

Mabin, A. (2018). Challenges of university–city relationships: Reflections from the university of the Witwatersrand and Johannesburg. In L. Bank, N. Cloete, & F. Van Schalkwayk (Eds.), Anchored in place: Rethinking the university and development in South Africa. Somerset West: .

Macupe, B. (2019, September 23). Unizulu's plans to build residences hampered because it does not own land. *Mail and Guardian*.

Martin, L. L., Smith, H., & Phillips, W. (2005). Bridging "town & gown" through innovative university-community partnerships. *The Innovation Journal: The Public Sector Innovation Journal, 10*(2), 1–16.

Matthews, Z. K. (1957). The University College of Fort Hare. *South African Outlook*, (April–May 1957), 1–35.

Maylam, P. (2005). Rhodes University: Colonialism, segregation and apartheid, 1904–1970. *African Sociological Review/Revue Africaine de Sociologie, 9*(1), 14–22.

Maylam, P. (2016). 'Oxford in the bush': The founding (and diminishing) ethos of Rhodes University. *African Historical Review, 48*(1), 21–35.

Miller, D., & Rivera, J. (2005). Town and gown: Understanding the past to improve the future. *International Journal of the Humanities, 3*(8), 215–224.

Moore, T. L. (2014). *Community-University engagement: A process for building democratic communities: Ashe Higher Education Report, 40: 2*. John Wiley & Sons.

Mosier, S. (2015). Does the gown help the town? Examining town–gown relationship influence on local environmental sustainability in the United States. *International Journal of Public Administration, 38*(11), 769–781.

Mulhearn, C., & Franco, M. (2018). If you build it will they come? The boom in purpose-built student accommodation in central Liverpool: Destudentification, studentification and the future of the city. *Local Economy, 33*(5), 477–495.

Ndimande, N. P. (2018). Student housing and the Slummification of the University of Zululand village in Kwadlangezwa, South Africa. *Human Geographies, 12*(1), 23–40.
Nkomo, M., & Sehoole, C. (2007). Rural-based universities in South Africa: Albatrosses or potential nodes for sustainable development? *International Journal of Sustainability in Higher Education, 8*(2), 234–246.
Ntema, J. (2021). Alice: Socio-spatial change in a small university town. In *South African urban change three decades after apartheid: Homes still apart?* Springer.
O'halloran, P. (2018). Contested space and citizenship in Grahamstown, South Africa. *Journal of Asian and African Studies, 53*(1), 20–33.
O'mara, M. P. (2012). Beyond town and gown: University economic engagement and the legacy of the urban crisis. *The Journal of Technology Transfer, 37*(2), 234–250.
Rashid, Y., Rashid, A., Warraich, M. A., Sabir, S. S., & Waseem, A. (2019). Case study method: A step-by-step guide for business researchers. *International Journal of Qualitative Methods, 18*, 1–13.
Revington, N. (2021). Post-studentification? promises and pitfalls of a near-campus urban intensification strategy. *Urban Studies, 59*(7), 1424–1442.
Reynolds, A. (2020). Geographies of purpose built student accommodation: Exclusivity, precarity and (im) mobility. *Geography Compass, 14*(11), 1–18.
Roller, M. R., & Lavrakas, P. J. (2015). *Applied qualitative research design: A total quality framework approach*. Guilford Publications.
Rowan, M. (2017). The illusion of broken windows theory: An ethnographic engagement with the theory that was not there. *William & Mary Policy Review, 9*, 1–50.
Saldaña, J. (2014). Coding and analysis strategies. In P. Leavy (Ed.), *The Oxford handbook of qualitative research*. Oxford University Press.
Schoch, A., Pienaar, A. J., Carlisle, J., Pring, H., Hope, A., Jenkins, E., & Morton, R. (1938). South Africa. *Journal of Comparative Legislation and International Law, 20*(2), 120–140.
Singh, O. (2019). Police station near Unizulu Torched Amid ongoing protests. *Times Live*.
Sladek, E. (2019). The transformative power of anchor institutions. *Metropolitan Universities, 30*(1), 1–4.
Smith, D. P. (2019). Studentification. In *The Wiley Blackwell encyclopedia of urban and regional studies* (pp. 1–3). Wiley Blackwell.
Smith, D., & Fox, M. (2019). *Studentification guide for North America: Delivering harmonious town and gown associations*. Loughborough University.
Smith, D. P., & Holt, L. (2007). Studentification and 'apprentice' gentrifiers within Britain's provincial towns and cities: Extending the meaning of gentrification. *Environment and Planning A, 39*(1), 142–161.
Stake, R. E. (1995). *The art of case study research*. Sage.
Thakrar, J. S. (2018). University–community engagement as place-making? A case of the University of Fort Hare and Alice. *Development Southern Africa, 35*(5), 673–688.
University of Zululand. (1968). *University of Zululand Calendar*. KwaDlangezwa.
University of Zululand. (1999). *Unizulu'99*. KwaDlangezwa.
University of Zululand. (2010). *Self-evaluation report of the Che*. KwaDlangezwa.
University of Zululand. (2020). *University of Zululand Council Statement: 12/03/20*.
Vandegrift, D., Lockshiss, A., & Lahr, M. (2012). Town versus gown: The effect of a college on housing prices and the tax base. *Growth and Change, 43*(2), 304–334.
Weideman, M. (2004). Who shaped South Africa's land reform policy? *Politikon, 31*(2), 219–238.
Weill, L. V. (2009). The President's role in cultivating positive town-gown relations. *Planning for Higher Education, 37*(4), 37.
Wicks, J. (2017). Unizulu bosses don't find cows amoosing. *Times Live*.
Wilson, J. Q., & Kelling, G. L. (1982). Broken windows. *Atlantic Monthly, 249*(3), 29–38.
Woldoff, R. A., & Weiss, K. G. (2018). Studentification and disorder in a college town. *City & Community, 17*(1), 259–275.

Chapter 9
Local Government Election 2021 Trends in Small Towns

Stephen Rule

9.1 Introduction

The Local Government elections of 1st November 2021 were characterised by two prominent trends. The first was the decline in support for the ruling African National Congress (ANC), on this occasion to below 50% for the first time since 1994. The downward trajectory has been continuous since the 2006 local government elections when the ANC received 62% of votes. The ANC's share declined to just under 60% in 2011, 55% in 2016 and 46% in 2021 (Table 9.1). The second trend was a decline in voter turnout from around 58% of registered voters in 2011 and 2016, to only 46% in 2021. This comprised a mere 35% of all potential voters. Similarly, support for the largest opposition party, the Democratic Alliance (DA) dropped from 26.3% in 2016 to 21.7% in 2021. The beneficiaries of the lower levels of support for both the ANC and DA, were a range of smaller parties. These were the Economic Freedom Fighters (EFF) 10%; Inkatha Freedom Party (IFP) 6%; Freedom Front (VF+) 2%; Action South Africa 2%; Patriotic Alliance (PA) 1%; and Independent candidates 2%. As encapsulated by Blackman and Dall (2022, p.346) "[A]round the country, voters were expressing their dismay with the ANC and DA, asserting instead that local is lekker".

This chapter investigates the local manifestation of the 2021 election results in four small towns, and the mitigating impact of local geographies on 'received' broader national political narratives and sentiment. The four towns were selected across four provinces, to enhance representativity. These were Sutherland, Philippolis, Cradock and Plettenberg Bay (Fig. 9.1), all of which were incorporated into larger local municipalities in the demarcation of 2000. Susan Rule and I took a week-long circular trip to these and other towns during the last week of October

S. Rule (✉)
Department of Geography and Environmental Studies, Stellenbosch University, Stellenbosch, South Africa

© The Author(s), under exclusive license to Springer Nature Switzerland AG 2023

R. Donaldson (ed.), *Socio-Spatial Small Town Dynamics in South Africa*, GeoJournal Library, https://doi.org/10.1007/978-3-031-37142-4_9

Table 9.1 Party support levels 2004–2021

Party	2004 National	2006 Local	2009 National	2011 Local	2014 National	2016 Local	2019 Local	2021 Local
ANC	68.6%	62.0%	65.0%	59.6%	62.2%	54.5%	57.5%	45.6%
DA	12.2%	15.7%	16.4%	23.3%	22.2%	26.3%	20.8%	21.7%
EFF	–	–	–	–	6.4%	8.0%	10.8%	10.3%
IFP	6.9%	7.3%	4.5%	3.5%	2.4%	4.2%	3.4%	5.6%
VF+	0.9%	0.9%	0.8%	0.5%	0.9%	0.7%	2.4%	2.3%
UDM	2.2%	1.0%	0.8%	0.6%	1.0%	0.6%	0.5%	0.5%
COPE	–	–	7.3%	2.0%	0.7%	0.4%	0.3%	0.0%
ACDP	1.6%	1.2%	0.8%	0.6%	0.6%	0.4%	0.8%	0.1%
PA	–	–	–	–	0.1%	0.1%	0.1%	0.9%
ATM	–	–	–	–	–	–	0.4%	0.6%
Action-SA	–	–	–	–	–	–	–	2.3%
Other	7.7%	12.0%	4.3%	9.9%	3.6%	6.2%	3.5%	10.1%
Total	100.0%	100.0%	100.0%	100.0%	100.0%	100.0%	100.0%	100.0%
Registered	20,674,923	21,054,954	23,181,997	23,654,347	25,388,082	26,333,353	26,756,649	26,204,579
% Turnout	76.7%	48.4%	77.3%	57.8%	73.5%	58.1%	66.1%	45.9%

(Source: IEC, 2021)

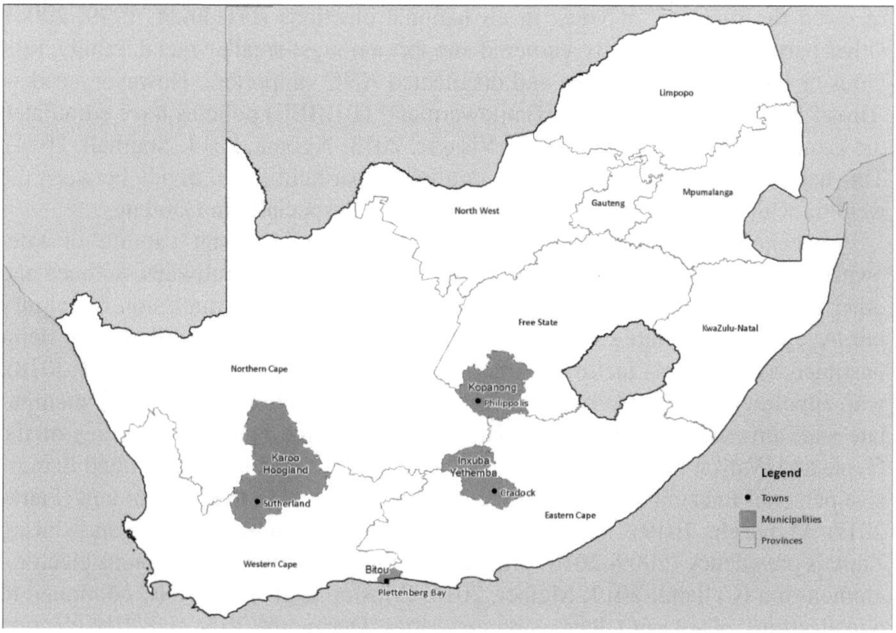

Fig. 9.1 Location of four small towns within their local municipalities. (Cartographer: Johann Fenske)

2021 and gathered perspectives on the troubled state of municipal government. Such was our engagement, as keenly interested voters, and participant observers, that local politicians were prepared to share deeply about their hopes and fears for their small towns. For these honest informants, I am truly grateful; their views are woven liberally into the fabric of this chapter.

9.2 Election Trends since 1994

Since the advent of universal franchise democracy in 1994, election trends have been closely monitored and analysed (Christopher, 1996; Khosa & Muthien, 1998; Ramutsindela, 1998; Lodge, 1999; Muthien, 1999; Olivier & Rule, 1999; Reynolds, 1999; Spence, 1999; Fox & Lemon, 2000; Rule, 2000; Ramutsindela, 2001; Mattes & Piombo, 2001; Maharaj & Ramutsindela, 2002; Mattes & Southall, 2004; Booysen, 2004; McLaughlin, 2007; Maharaj, 2009; Lemon, 2009; Russon, 2011; Netswera, 2012; Seethal, 2012; Southall & Schulz-Herzenberg, 2014; Booysen, 2015; Rule, 2018; Kotze & Bohler-Muller, 2020; Roberts et al., 2020; Sadie & Patel, 2020; Dhawraj et al., 2021). Race and ethnic identity have played a major role in determining voting patterns, and Black African voters, who comprise more than three-quarters of the electorate, have tended to support the ANC, which has hitherto received the majority of votes in all national elections (Friedman, 1999, 2004). Other parties have generally garnered support amongst smaller racial, ethnic, religious or other interest groups, and disaffected ANC supporters. However, strident 'Broad-Based Black Economic Empowerment' (BBBEE) policies have stimulated the growth of a black middle class (Visagie, 2013; Ngoma, 2014; Southall, 2014). This has served gradually to blur the dominant apartheid social divide between the wealthy white/Indian and poor black components, especially in Gauteng.

Associated with BBBEE has been the widespread and rampant 'capture' of state-owned enterprises in the electricity, water, security, airways, railways, defence and other sectors of the economy (Zondo, 2022). The Zondo Commission of Inquiry into Allegations of State Capture, Corruption and Fraud in the Public Sector demonstrated how this was facilitated inter alia by President Jacob Zuma (2009–2018), who appointed officials to political and managerial positions, who would manipulate state procurement processes that would favour the business activities of the Gupta and Watson families and others (Zondo, 2022). This, in turn, resulted in massive personal financial benefits for his corrupt allies within the government (Pauw, 2017; Myburgh, 2019). Although the large-scale corruption before and during Zuma's presidency (2009–2018) provided opposition parties with copious electoral ammunition (Calland, 2013; Melber, 2014; Prinsloo, 2014), the ANC continued to win elections, albeit with decreasing majorities. During the 2014 and 2016 election campaigns, the ANC was frequently accused of using government resources for its partisan electioneering activities (Butler, 2014; Montsho, 2016). The DA profiled its relative success in local governance in the Western Cape, and increasingly in the major metropoles of in Gauteng (Johannesburg, Tshwane, Ekurhuleni) and the Eastern Cape (Nelson Mandela Bay). The EFF won a case in the Constitutional Court, which ordered President Zuma to refund some of the costs of renovations and additions to his private residence at Nkandla. In addition, between elections, voters have protested frequently (Seethal, 2012; Alexander et al., 2018) owing to anger over corruption, socio-economic inequalities, poor municipal service delivery and dysfunctionality in one-third of the country's municipalities (Nel, 2016). Runciman et al. (2016) estimated that there were 71,000 protests between 1997 and 2013.

Election results have continually reflected core-periphery differences in poverty, wealth, income and assets that persist in the democratic era (Noble & Wright, 2013). Provinces with greater proportions of rural residents (Mpumalanga, Limpopo, Eastern Cape) have persistently supported the ANC more strongly than provinces with the highest urban components (KwaZulu-Natal, Gauteng, Western Cape). Similar relative patterns have recurred between the poorer and wealthier settlements in cities. In Johannesburg for example, 2016 local government voters in Ward 15 (Senaoane, Soweto) were most supportive (79.8%) of the ANC; those in Ward 126 (Weltevreden Park, Randburg) were most supportive (87.3%) of the DA; and the EFF performed best (31.4%) in Ward 111 (Ivory Park North), a low income partially informal settlement.

In the most recent decade, increasing proportions of the electorate have begun to identify as politically independent and non-partisan (Schulz-Herzenberg, 2013). A telephone survey of 3905 voters living in Johannesburg, Tshwane, eThekwini, Cape

Town, and Nelson Mandela Bay after the 2021 elections determined "declining party loyalties and growing fluidity with respect to voter choices" (Paret & Bekker, 2021, p. 6). Less than half (49%) of ANC 2016 voters repeated their choice in 2021, as did only 55% of those who voted for the DA in 2016. Similarly, only 35% of EFF 2016 voters voted again for the EFF in 2021. Paret and Bekker (2021, p. 6) further found that "[M]ore than one quarter of those who voted in 2016 chose to abstain from voting in 2021. The ANC lost 33% of its voters to abstention, the EFF lost a remarkable 41%, the DA lost 28%, and parties beyond the big three lost 35%.

9.3 Local Government Dysfunctionality

The pre-democracy system of local government originated during the colonial era and entailed progressively more stringent racial segregation of residential areas. Only in the 1980s did Indian and black African households begin to re-settle in central Johannesburg and the old central suburbs of other cities, when some of the apartheid legislation was relaxed. The system had been characterised by heterogeneous segregated municipalities that varied by magnitude, area of jurisdiction, powers, functions, and sources of revenue. The apartheid boundaries, lack of uniformity, and mismanagement of financial resources (Worrall, 1971; Binza, 2005) were then addressed by the Local Government Municipal Systems Act (South Africa, 2000). The Act was intended to "provide for the core principles, mechanisms and processes that are necessary to enable municipalities to move progressively towards the social and economic upliftment of local communities and ensure universal access to essential services that are affordable to all" South Africa, 2000, 1). Aspirations were to adhere to the principles of efficiency and effectiveness of municipal leadership in the New Partnership for Development (NEPAD), and the Commonwealth Roundtable on Decentralisation and Democratisation for Senior Local Government Policy makers in Africa. These included democratisation and decentralisation, fundamental human rights, the rule of law, the independence of the judiciary, just and honest local government, empowerment of people at local government level, unrestricted political activity, free and fair local government elections, and allocation of the necessary resources and technical expertise by central governments and the international donor community (Binza, 2010). However, a tragic and unintended consequence of the amalgamation of multiple small towns (834 segregated municipalities) into 234 post-apartheid municipalities was to deprive many existing small towns of their "identity or voice". The positive intentions behind eradicating lingering apartheid geographies thus undermined the downgraded status of these towns, and their subsequent impaired capacity to facilitate local economic development (Hoogendoorn & Nel, 2012, 23). A longer-term consequence has been to dissipate public accountability and render the majority of municipalities partly or wholly dysfunctional, close to bankruptcy and unsurprisingly, unable to comply with the requirements of the Auditor-General (South Africa, 2022).

In spite of the immensity of the challenges, people are living there, improvising, volunteering, and seeking solidarities. The evidence presented, comprising cameos on the 2021 political context and political dynamics in four towns, suggests perseverance and resilience. The towns were selected purposively in four adjoining provinces, as exemplars of many others situated across the rural hinterlands of the country.

9.4 Sutherland

Prominently branded as the site of the South African Astronomical Observatory, Sutherland is situated in Northern Cape, north of the Roggeveldberge range and 350 km north-east of Cape Town. With a population of 2836 in 2011 (City Population, 2011), Sutherland is the smallest of three towns in the Karoo Hoogland Local Municipality, the others being Williston and Fraserburg. It was established as a church and market town for sheep farming in 1855, and the town population comprises 78% coloured people, 13% whites, and 8% black Africans.[1] Linguistically is almost homogeneous at a level of 99% Afrikaans-speaking (StatsSA, 2016). In the local municipality (LM) as a whole, about half (46%) of adults are in employment and the mean household income is R29,400 (StatsSA, 2011). Owing to the tropospheric stability of the Karoo, the South African Large Telescope (SALT) was built near Sutherland. This serves as a global attraction for astronomers and other scientists and curious visitors (Ingle, 2010). The valuable research facilitated by SALT is published in journals such as the *Monthly Notices of the Royal Astronomical Society* on circumstellar nebulae, supernova remnants, solar oscillations and optical spectroscopy (see SALT, 2021; Gvaramadze et al., 2018). Local benefit is highlighted in accommodation reservations websites, which feature dozens of options in and around the town: bed and breakfast houses, hotels, and self-catering cottages. Restaurants and other star-gazing related businesses cluster to take advantage of demand, thus creating local employment, although severely curtailed during the pandemic.

The total population of Karoo Hoogland LM grew slowly from 12,588 in 2011 to 13,069 in 2016. Observation suggests that the 2022 census will reflect a similar trend. Just over half (R31-million, i.e. 54%) of the municipal budget of R57-million is generated locally from payments made by residents for water, electricity, rates, licences, fines, and from interest and investments (Municipalmoney, 2022). The balance (R26-million, i.e. 46%) comprises financial transfers from the 'Equitable Share' of taxes, and grants from national and provincial government.

Karoo Hoogland LM comprises six wards, an increase from four after the 2016 local government elections. Three wards include parts of Williston and Amandelboom

[1] The ethnic categories African, coloured, white and Indian are still used in official statistics in South Africa.

and the surrounding farms (Ward 1 east; Ward 3 north; Ward 5 west); Ward 2 is the Fraserburg area; Ward 4 is urban Sutherland; Ward 6 is the surrounding hinterland of Sutherland in all directions.

Mr. Jan Jacobs (Fig. 9.2), councillor and DA candidate in Ward 4 agreed to an interview and his commitment to public service was immediately evident. He grew up in Sutherland and held jobs in the local health clinic, during which he was exposed to the daily struggles and issues of local residents. He then worked at a supermarket in Cape Town and returned to care for his mother in 2014. He decided to contest the 2016 election with a view to improving life conditions for residents and was elected as a PR councillor. His view is that being proactive rather than passive is the way to achieve improvements [*"as jy 'n glimlag op een mens se gesig per dag kan sit, dan het jy wondere verrig"*].

Mr. Jacobs' experience of the town is that perspectives have changed and are continually changing and that there is a 'hunger for change' amongst residents. His cousin was the ANC candidate in the same ward, illustrative of an activist family. Mr. Jacobs expressed despondency about the town's high level of unemployment, the shortage of two-bedroom houses, the existence of numerous potholed roads, high rates of tuberculosis and Covid-19 infections and the lingering 60% prevalence

Fig. 9.2 Mr. Jan Jacobs, DA candidate, Ward 4, Karoo Hoogland, 2022. (Photo: by author)

of pit toilets. A glimmer of hope is emerging as green energy pursuits expand, with a large proportion of local jobs being related to wind farms in the district. Hinting at corrupt municipal practices, he remarked that municipal council meetings do not keep adequate attendance records and that money "disappears". For example, the ruling council had misused funding designated for toilets, and spent it on water. His impression was that voters aged over 40 years tended to support either the ANC or DA, and younger voters were opting for the EFF, although many of them were migrating to the Western Cape, in search of work. His perception was also that the ANC was losing support to independent candidates. Nevertheless, the municipality performed relatively well to receive a verdict financially unqualified with findings in the report of the Auditor-General 2020–21. Over the previous 4 years, the audits were financially unqualified with findings in 2 years and qualified with findings the other 2 years (South Africa, 2022).

The 2021 local government elections in Karoo Hoogland were highly contested by the dominant ANC, the main opposition DA, and eight other political parties (Fig. 9.3). At 62% in both 2016 and 2021, voter turnout was well above the national average (46%) and even the Northern Cape average (53%), indicative of meaningful investment in the outcome. As in 2016, the ANC and DA retained their edge over smaller contestants but with far less dominance, thus reflecting the national trend.

The ANC's share of the vote declined from 54% to 40%, signifying a rejection of decades of poor governance (Fig. 9.4). Local candidates thus bore the brunt of punishment being meted out, probably less for local municipal inefficiencies than on the party's national leadership for inept handling of the misdemeanours of former President Zuma and his acolytes.

Fig. 9.3 Election posters in Sutherland (Karoo Hoogland), October 2021. (Photos: by author)

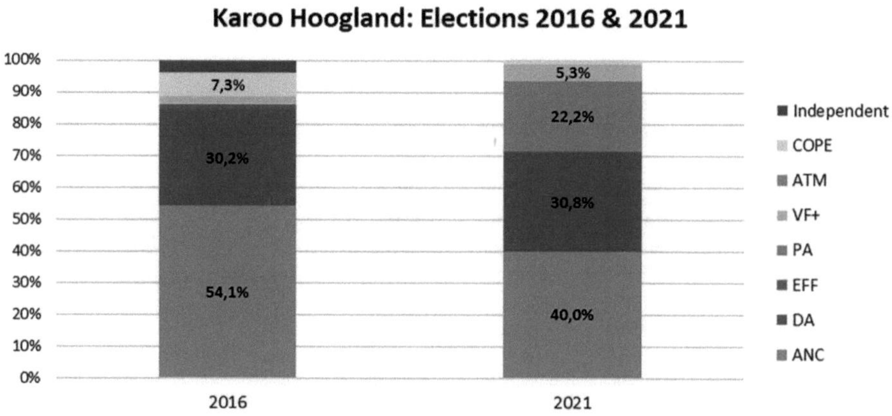

Fig. 9.4 Karoo Hoogland local government results, 2016 and 2021. (Source: Derived from IEC, 2021)

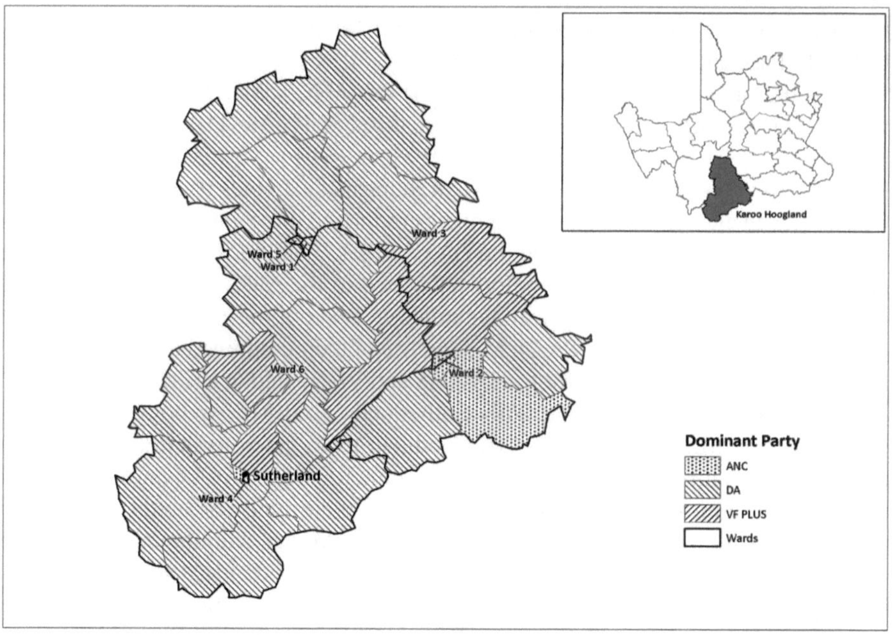

Fig. 9.5 Dominant party in each ward, Karoo Hoogland, 2022. (Cartographer: Johann Fenske;, data derived from IEC, 2021)

The DA managed to retain the same level of support as received in 2016 (30%), attracting the loyalty of coloured and white support base living on farms or in townships and suburbs (Fig. 9.5). The greatest change was the massive growth in support for the Patriotic Alliance (led nationally by Mr. Gayton McKenzie), at 22% across the municipality. COPE support declined from 7% to only 1%, and the VF+ doubled

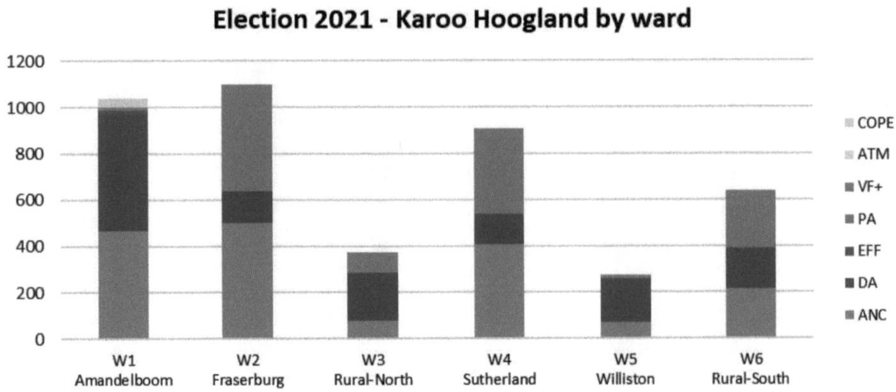

Fig. 9.6 Number of votes in each ward, Karoo Hoogland, 2022. (Source: Derived from IEC, 2021)

its support to 5%. The PA's success was arguably attributable to clear and consistent messaging (Sussman, 2021) targeted at meeting the needs in specific coloured-majority communities, such as Sutherland and Beaufort West (22%), as well as Eldorado Park and Reiger Park in Gauteng.

In Sutherland's Ward 4 itself (Fig. 9.6), the ANC's Ms. Sandra Muller was elected as councillor with 408 votes (45%), ahead of PA Mr. Dawid Baadjies with 340 votes (37%), DA Mr. Jan Jacobs 129 votes (14%), VF+ Mr. Riaan van der Merwe's 30 votes (3%) and EFF Mr. Alian Stuurman's 3 votes (0.3%).

The eleven elected LM council members comprised three wards councillors each for the ANC (wards 2, 4 and 6) and the DA (wards 1, 3 and 5), supplemented by five proportional representative councillors, allocated to the ANC (2), PA (2) and VF+ (1) respectively. A remarkable local dynamic is that six of the eleven councillors are based in Sutherland (ANC 4, PA 2). Making use of their combined majority over the ANC, the councillors elected for the DA, PA and VF+ opted to collaborate and took control of the municipality. The PA's Sutherland-based Mr. Anthony Mietas was elected Mayor. Mr. Jacobs (losing DA Candidate) was gracious in defeat and positive about the current functionality of the Council, suggestive of an amiable and collaborative atmosphere with political opponents striving for better services to their communities, despite his cousin Mrs. Sandra Muller winning Ward 4 for the ANC. This bodes well for Sutherland, and the Mayor confidently invokes the town's identity as an "Astronomy and Tourism Node", home to the South African Large Telescope (SALT) and proximity to the Square Kilometre Array (SKA). The Mayor asserts that "The tourism industry is ripe to be the largest contributor to the economy of South Africa. Tourism is the second biggest sector in terms of job creation ranking only below agriculture in the country's economy. The 30-year tourism Master Plan is undertaken to boost tourism by identifying and exploring products that will attract tourists into Carnarvon Sutherland Corridor and the Northern Cape Province" (Karoo Hoogland IDP, 2022–27, p. 9).

9.5 Philippolis

Located in the Free State along the R717, and 175 km south-west of the provincial capital Mangaung, Philippolis (population 3648 in 2011) is one of the seven small towns in the Kopanong Local Municipality. The name derives from John Philip of the London Missionary Society, which established a station there in 1823 amongst the nomadic Griqua population. Almost two-thirds (65%) of the residents of Philippolis are black Africans, 27% are coloured and 6% are white (City Population, 2011), and the most spoken home languages are Afrikaans (41%), isiXhosa (29%) and Sesotho (26%). The Kopanong LM as a whole has a total population of 49,999 (49,171 in 2011), and a higher proportion Sesotho-speakers (41%) and lower proportions of the Afrikaans (33%) and isiXhosa (21%) groups (StatsSA, 2016). Just over a third (36%) of adults are in employment and the mean household income is R29,400 (StatsSA, 2011). As with other Karoo towns such as Calvinia, De Rust, Fouriesburg, Graaff-Reinet, Greyton, Prince Albert and Richmond (Ingle, 2009; Donaldson & Marais, 2012; Donaldson, 2017), counter-urbanising middle class people from the large cities have recognised the local charm and potential for a quieter life. Gentrification has accordingly boosted tourism, local property values and local employment opportunities.

Unfortunately, municipal governance has not adapted appropriately. The LM had a total revenue of R185-million in 2010/11, of which only 36% was locally generated and R118-million (64%) comprised government grants and subsidies (Kopanong Local Municipality, 2011). The municipal audit of 2017–18 (South Africa, 2019) showed a net loss of R84, 7-million and that current liabilities exceeded current assets by R367, 7-million. Although the municipality had deducted R58-million worth of tax, pension and medical aid contributions from employees' salaries, they had not paid these over to the relevant third parties (R40-million in 2017). A municipal report had not been compiled in time for the Auditor-General's 2020–21 audit, but in all four previous 4 years, the AG's verdict was 'qualified with findings' (South Africa, 2022).

Kopanong LM comprises nine wards. Ward 4 includes urban Philippolis and Poding-Tse-Rolo and Bergmanshoogte and the farms to the north, west and south of the town. Six other wards each include a small town and surrounding rural areas (Ward 1: Reddersburg; Ward 2: Trompsburg; Ward 3: Bethulie; Ward 5: Springfontein; Ward 6: Fauresmith; Ward 8: Edenburg). Two others are predominantly rural in the north (Ward 7) and south-east (Ward 9) of the LM.

The ANC candidate in Phillippolis (Kopanong Ward 4), Mr. Vuyisile Jonas agreed to an interview and indicated that he had attended primary school in Philippolis and high school in the former Ciskei homeland. He was a clinic assistant and a municipal policeman and had worked with Professor Doreen Atkinson (Karoo Foundation) in a municipal issues NGO. He was elected as an ANC councillor in 2000, and as Mayor until 2006. Mr. Jonas opened a local tavern and club, which resulted in some friction with local residents, and invested in local housing. He was re-elected in the 2006 elections but expelled in 2009 because he sided with the

Thabo Mbeki faction while other councillors supported Jacob Zuma for president. For a while he lived in the Eastern Cape and was a member of COPE, but returned to Kopanong, in 2020, when the ANC approached him to stand as a candidate in 2021. His view was that during his absence, technical incompetence and lack of skills have hindered delivery of services. Employment is the town's primary need but there are "no willing investors" apart from those with "poultry projects etc., which provide no long-term benefits to workers." Expanding on the contemporary woes of the municipality, Mr. Jonas indicated that the "account of the municipality is attached, because money is owed to Eskom and Bloemwater. Bloemwater closes off water at 12pm daily and certain areas are deprived in afternoons. Municipal workers are not working. The 'equitable share' is inaccessible and R1-billion is owed to creditors' '. He pointed out that "most paying ratepayers are in white areas' ' and much of the rest of the population "lives below the poverty line." Most local employment is in the public sector (police, clinic, traffic, municipality, school) and farm workers are poorly paid, sometimes being "replaced by short-term contractors." He recalled that he had participated in the process of amalgamating the previous municipalities into the unified Kopanong LM, and that "only Philippolis, Reddersburg and Gariep Dam were financially viable and effective at the time (2000)". Regarding other political parties, he said the 'opposition only opposes" and that the "EFF attracts about 10% of the youth", owing to their "anger and frustration and the view that "Mandela did nothing for me". His perception was that the DA obtains most of its votes from white and a few coloured and black voters. Owing to good relationships with their farmer employers some black voters support the DA, taking the view that "*die baas is altyd reg*". He doubted that the VF+ would receive much support. Overall, "most residents are poor, and they still have sympathy with the ANC. The problem lies with the national leadership". Regarding the performance of municipal officials, he averred that "locals must take ownership of their accounts. We cannot blame everything on incompetent staff". He said that the requirements for municipal posts are legislated and therefore that "the correct cadres must be deployed". The ANC should "look for the person who is able to do the job".

Ms. Estelle Noordman, DA candidate in Philippolis, told us that she had visited the town frequently before settling there in 2016 when she finished her job as a Bloemfontein-based medical sales representative. She said that municipal inefficiencies had motivated the local Chamber of Commerce to intervene. Her personal frustration with poor maintenance and delivery of municipal services in the town had prompted her activist participation in the Ratepayers' Association (RA). On one occasion she reported poor water pressure to councillors at the LM offices in Trompsburg. Ms. Noordman accompanied them to the local reservoir and sewage works, where it became evident that water supplies had been illegally diverted and sewerage reticulation had been disrupted, allegedly in order to procure work for particular local service providers. Accordingly, service providers were appointed to make the necessary repairs, but were unable to do this adequately. Her view was that low levels of municipal management efficiency were intolerable (*Tot hier toe en nie verder nie! Asseblief mense, hierdie goed moet regkom*).

Professor Doreen Atkinson confirmed that Ms. Noordman, in collaboration with local administrative and technical officials, had personally organised several "hands-on" interventions to address local water and sewerage challenges, using funds donated by the RA and the principal of a local school from which water was being "stolen". Ms. Noordman said that residents were unable to pay their rates and services accounts because the computer system was inoperative, and officials were not available because they were not being paid and were on strike. She alleged that officials misunderstood the processes of debiting and crediting household accounts. Although the municipality is constitutionally obliged to provide 30% of water to municipalities, this has proven unsustainable for Bloemwater, the regional water authority to which nine towns owed R461-million by October 2021. Only Jagersfontein was able to function following a bailout of R6-million from mines in the area. Ms. Noordman attributed increasing property values in Philippolis to the efforts of the RA but lamented that the RA was not entitled to engage directly with Bloemwater. Her view was that were each town in Kopanong to form a RA, "Bloemwater may be able to deal with the combined entity".

Prof Atkinson said that the November 2021 elections would be the "first time people would be voting for a council for which there is no operational municipality…… no cash, no workers" and she called for the national Department of Cooperative Governance and Traditional Affairs to manage the situation. She said that workers were striking, and no one was being paid, including the councillors. Prof Atkinson alleged that Kopanong IDP officials had indicated that 300 houses and a football stadium worth R2-million were to be constructed and that because impropriety in the award of tenders was suspected, the Hawks visited in 2020 to investigate. Pertaining to election dynamics, Prof Atkinson referred to the existence of three ANC factions in Poding. Nominees for the Ward 4 seat had been existing councillor Mr. Momo Jan and aspirant Mr. Vuyisile Jonas, and there were allegations of manipulation of vote counts in the preliminary election for a candidate.

The Kopanong LM elections in 2021 were contested by ward level candidates representing five main political parties and an independent candidate, as well as two other parties for PR seats only. Less than half (49%) of registered voters participated, but nonetheless marginally more than the Free State provincial average of 45%. Grossly low municipal performance levels would have discouraged engagement owing to black African voter perceptions of the lack of an alternative to the hugely dominant incumbent ANC. In the event, there was little substantial change in the results between 2016 and 2021. In line with the national trend, the ANC support level declined by just over 3% (Fig. 9.7), from slightly more than two-thirds (68.1%) to slightly less than two-thirds (64.8%). The support was sufficiently well spread across the LM to enable the ANC to win the largest number of votes in every one of the nine wards.

The DA achieved a marginally lower proportion of votes (declining from 17.2% to 16.5%). The EFF received a 50% boost from 6.3% to 9.3%, exploiting dissatisfaction amongst former ANC loyalists. The VF+ jumped from 2.3% to 5.7%, presumably siphoning off potential growth in white voter support for the DA.

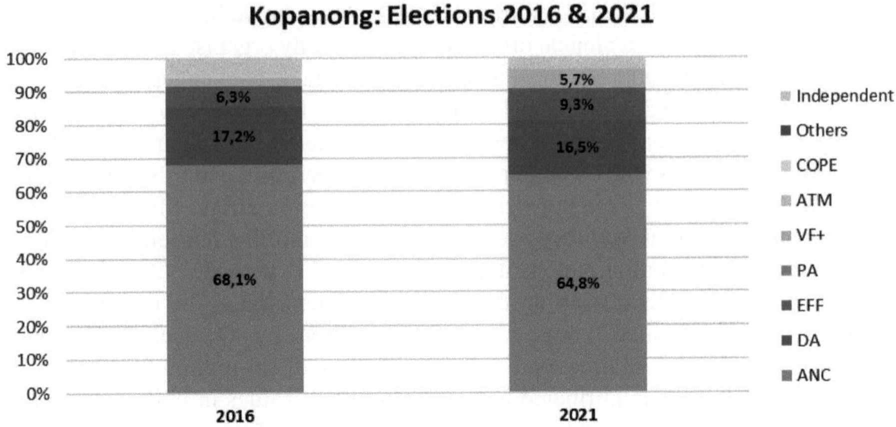

Fig. 9.7 Kopanong local government results, 2016 and 2021. (Source: Derived from IEC, 2021)

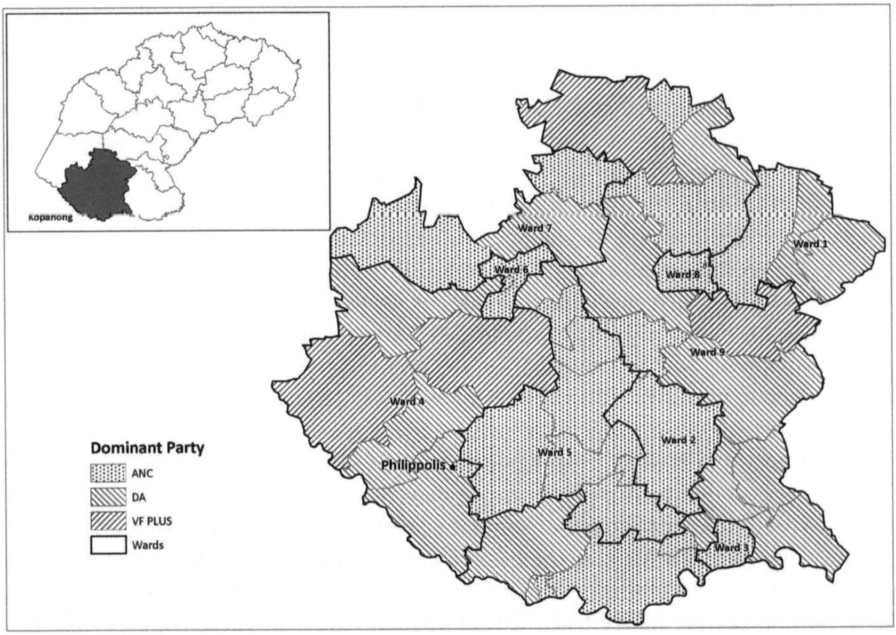

Fig. 9.8 Dominant party in each ward, Kopanong, 2021. (Cartographer: Johann Fenske; data derived from IEC, 2021)

In only two (wards 4 and 9) of the nine, was ANC support lower than 50% (Figs. 9.8 and 9.9). In Philippolis (Ward 4 of Kopanong LM), the ANC's Mr. Vuyisile Jonas was elected as councillor with 620 votes (49%), ahead of DA Ms. Estelle Noordman with 449 votes (36%), EFF Mr. Tshepang Mokone 146 votes (12%), and VF+ Mr. Pieter du Toit 42 votes (3%).

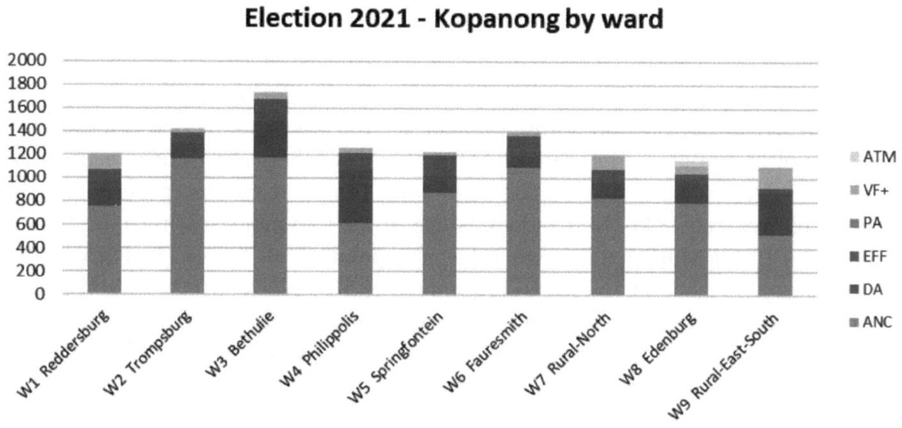

Fig. 9.9 Number of votes in each ward, Kopanong, 2021. (Source: Derived from IEC, 2021)

Ms. Noordman was nevertheless also elected to the Council by virtue of being a PR candidate for the DA, as well as a ward candidate. The seventeen elected LM council members comprised the eight ANC ward councillors, supplemented by three ANC PR councillors. The DA won three PR council seats, the EFF two, and the VF+ one. The ANC and DA activists in Philippolis, whom we interviewed before the election, were both thus able to continue their local political activities in formal political roles.

An update in June 2022, from Professor Atkinson painted a bleak picture. Her view was that although the Kopanong Municipal Manager is "doing her best…., the systems are not being put in place, …. municipal accounts are not being sent out, so revenue is probably zero, other than the equitable share from Provincial Government". She confirmed that the municipality "has no cash flow, and cannot pays its workers, who are again on strike". Nevertheless, a spirit of community awareness and preparedness to pitch-in is strong. Residents are intervening and managing refuse disposal amongst themselves in some neighbourhoods. Bloemwater is continuing to deliver a constitutionally required quantity of water for a few hours per day to each town, sometimes regulated, otherwise irregular, depending on local capacity. She said, "I suspect the challenges are just way too much for the municipality" despite the availability of guidance from CoGTA, and its request for the Mayor to meet with her (Professor Atkinson). "It is basically drifting like a rudderless boat". Prof Atkinson lamented that a donation which she had sourced for repairs to the community hall had been blocked. Similarly Councillor Noordman had purchased water tanks for houses that do not have them, but the Council declined to connect these to the mains for unspecified technical reasons. Her view is that "the DA is by far the most active and responsive agency".

9.6 Cradock

Situated in the Eastern Cape along the N10, on the west bank of the Great Fish (Inxuba) River, and 250 km north of Gqeberha, the town of Cradock emerged as a colonial military outpost in 1812, after the fourth war between European colonists and Xhosa residents. It was named after John Francis Cradock, governor or the Cape Colony from 1811–14. Cradock is one of the two major towns in the Inxuba Yethemba Local Municipality, which has a population of 70,494 (StatsSA, 2016) (modest growth from 65,560 in 2011) comprising 56% black Africans, almost all of whom are home language isiXhosa-speakers, while the coloured (36%) and white (8%) groups are predominantly Afrikaans-speakers. At a level of 39,7% employment, the mean annual household income was a low R29,400 in 2011. The town of Cradock itself accommodates 36,671 people, of whom 62% are black Africans, 25% coloured and 12% white (City Population, 2011).

The local economy grew with wool booms of the 1850s and 1870s: fencing and other investments capitalised the sheep farms, and dams and weirs proliferated, also supporting agriculture along the valley (Mabin, 1984, 2021; Beinart, 2008: esp. 173–4) and the town became a strategically located commercial centre. By that time, a ‘location’ existed at the south end of the town, to which were relegated most people other than white – especially isiXhosa speakers. Between the township and the colonial town centre lay a less determinate section in which lived many of the coloured and Asian-descended population, along with some of the less well-off white people (none of these categories being legally distinguished as they were after 1950). ‘Segregation and apartheid came to Cradock’ (in Jeffrey Butler’s 2017 title) in numerous and contested phases, culminating in the destruction of the old location, removals of ‘disqualified’ people from the town, and by degrees new housing schemes that grew into Lingelihle township (‘a good effort’). In the 1970s when coloured people were removed from the town, they went to Michausdal township across the national road from Lingelihle, with a mix of housing types typical of many such areas.

More than three-quarters (79%, i.e. R193-million) of the municipal budget of R246-million was generated locally from payments made by residents for water, electricity, rates, licenses, fines, and from interest and investments (Municipalmoney, 2022). The balance of R52-million (21%) comprised financial transfers from the ‘Equitable Share’ of taxes, and grants from national and provincial government. For each of the last 5 years, the Auditor-General has concluded ‘financially unqualified with findings’ for Inxuba Yethemba LM (South Africa, 2022).

Inxuba Yethemba LM consists of nine wards, five of which include parts of urban Cradock. These are Lingelihle south (Ward 1), Lingelihle east (Ward 2) and Lingelihle north-west (Ward 3), as well as Michausdal (Ward 4) and Cradock central (Ward 5). The other four are the rural hinterland of Cradock (Ward 6), eastern Midros and the rural hinterland of Middelburg (Ward 7), northern Kwanzoname and western Midros (Ward 8) and Middelburg and southern Kwanzoname (Ward 9).

Three election candidates (two ANC, one DA) agreed to speak to us. The ANC candidate in Ward 2, Mr. Monwabisi Rasmeni indicated that he had been fighting for two decades on behalf of unemployed young people in Cradock and in the Western Cape. He lamented the tendency for activists to lose their fervour once elected to political positions. This angers communities and discourages voting. He said that his campaigning had made him realise that people ``don't want the ANC anymore" but his view is that "the party is not at fault, it is corrupt people".

Mr. Andile Dakuse, ANC candidate in Ward 3 said that he has been involved in politics since 1965 when he was a teenager. He was inspired by his family's participation in the liberation struggle since the 1920s, and particularly by his relative Matthew Goniwe, one of the 'Cradock Four' who died 1983. He objects strongly to the ANC's policy of deploying people who lack the technical skills and capacity for municipal jobs, but he fears that many officials "don't want to hear the truth" about the ineffectiveness of employing friends.

Neither of these two ANC candidates had previously stood for election but had been involved with the Cradock Concerned Residents (CCR) civic organisation which arranged an anti-corruption march in the town after allegations of municipal mismanagement, cattle theft and sexual abuse. This resulted in investigations by the Hawks and CoGTA. Recommendations from the resultant report had yet to be implemented.

They regretted the apparent lack of effort by previous councils to attract external investors and to stimulate the local economy. There is potential for expansion of the dairy industry and for historical tourism, related to the legacies of Olive Schreiner and the Cradock Four (Fort Calata, Matthew Goniwe, Sicelo Mhlauli, Sparrow Mkhonto). The Schreiner museum is well managed, and the Cradock Four monument is spectacular but is incomplete and requires better maintenance. A large proportion of local jobs are agricultural, and many farmworkers live in Lingelihle. The condition of existing buildings and infrastructure is deteriorating.

They expressed concern about misuse of the R250-million annual municipal budget, improper procedures in the appointment of the municipal CFO, nepotism in the award of a construction project, and inefficiencies in the water supply and sewerage reticulation systems. They declined offers of R500,000 to stop the protest march. Their perception was that the record and associated credibility of the CCR would mobilise otherwise disgruntled voters to participate in the election because they had been nominated "by the community" rather than imposed by higher echelons in the party. They acknowledged the strength of the DA "in town" but did not anticipate much support for smaller parties except the EFF (Fig. 9.16) amongst the youth, although their candidate was "not a local guy", who had "only come for the election". They were optimistic about the potential for multi-party collaboration because "all parties want to see improvement.. [and]… share common interest in development and … our people getting employed …. and repair of potholes". It was essential for the Council to change and that the mayor should "be a good man". Their choice would be Thembi Mbotya, number one on the party's PR list, however a contingent of their colleagues favoured retention of the serving Mayor. "The provincial ANC knows about the corruption… some are involved". They alleged a financial "kickback" for a council official from a plumbing contractor, but that the investigation and prosecution process had been delayed until after the election.

The DA candidate for Ward 5, Ms. Rika Featherstonehaugh, had lived in Cradock for 15 years and represented the area in the Council since the 2016 elections. She explained the criteria for selection as a candidate, including a minimum education level of matric, and being involved in the community. She said that it had been "really hard work for the last 5 years…. people phone me 24/7". She complained of massive corruption and fleecing of residents by the Council and officials, which impacts on service delivery. Ms. Featherstonehaugh mentioned examples of R380-million being paid for the resealing of a street, to a Pretoria-based contractor, poor conditions at the landfill site, potholes, lack of street lights, illegal dumping, and the victimisation and suspension of an internal auditor. The Council owed Eskom R36-million 4 years ago, a debt that has increased to R230-million. Another tender had been awarded for repairs to two streets in Middelburg and three streets in Cradock. A case was lodged with the Public Protector and all complaints had been found to be justified. Implicated officials and councillors were forced to resign, and a new mayor Ms. Noncedo Zonke was elected. Additionally, Ms. Featherstonehaugh indicated that water supply infrastructure was not being maintained, sewerage pipes were blocked or broken, 500 street lights had not been working for more than 4 years. The municipality can find or afford neither a qualified electrical engineer, nor a mechanic with experience on petrol and diesel vehicles. Vehicles were not available for pothole repairs. Municipal revenue is inadequate for all that needs to be accomplished.

Ms. Featherstonehaugh said that residents' interest in the election was "very low" because they were despondent and "gatvol of politics" that was "all talk, no action". Nonetheless, she was confident of a better performance by the DA in 2021, and an improvement from the current 7 out of 18 councillors. Ms. Featherstonehaugh anticipated good results in the urban centres of Cradock and Middelburg and the surrounding farms and declined support for the ANC in their township strongholds. She said, "Despite voter apathy – this is the one election where there could be a turnaround. In nearby Jeffrey's Bay, there is an unbelievable turnaround in finances and service delivery. They are on another level. The best run municipality in the Eastern Cape. Where the DA gets in, they do a good job. They make mistakes, but they make some progress".

Ms. Featherstonehaugh's analysis was that there was a split in loyalty between the eleven ANC councillors owing to different views on corruption and the personal benefits being enjoyed by some. The principled ANC councillors who oppose corruption are sidelined and accused of working with the DA. She expected that support for the ATM and independent candidates would be at the expense of the ANC and was concerned that the PA and VF+ respectively would reduce coloured and white voter support for the DA.

Potential for economic growth existed in agriculture and related the associated engineering, tractor, irrigation equipment and seed industries. Dairy farming productivity in the region is the second highest in the country but "dairy farmers have to spend vast amounts of money to get the water from the furrows …[but]… ecoli is found in the water, which threatens a sophisticated industry involving rotary milking machines that milk 2,000 to 3,000 cattle several times a day per farm. The produce is distributed as far as Bloemfontein, Coega and Humansdorp. The town has 3 large supermarkets, 2 wholesalers, 2 liquor, and several clothing stores and banks. Tourism used to be "very strong" but it has "taken a knock. Previously, 20% of guests had been from overseas, but this proportion has declined and of the 50 guesthouses in Cradock, now about 30 are left. Business at Ms Featherstonehaugh's own guesthouse had totally disappeared during the Covid-19 pandemic but in recent weeks it had served as a quarantine base at 50% of normal cost, for Covid positive guests awaiting further test results. Festivals for rugby matches and writers, and the Cradock Show had been cancelled. She expressed regret that Matthew Goniwe's widow, Ms Nyameka Goniwe, speaker of the LM had died in 2020. "She was pushing back against corruption. She had a voice". Ms. Featherstonehaugh lamented the LM being "on brink of financial collapse" saying that municipal salaries were paid before the scheduled debit orders for SARS, which amounted to over R9-million. "After salaries had been paid there was just R53 left in the municipal account". Although the LM received an unqualified audit, there were "pages and pages of questions". In her ward, there were 3222 registered voters and she expect about 2000 votes to be cast. Registrations had increased by more than 200 since the 2016 elections in spite of deaths and out-migrations, but there were "still 400 people in the ward who have not registered correctly". She was confident in the integrity of the election management and recalled that in 2016 "counting agents went into town hall at 7am and remained until 9pm when voting closed… They counted till 12 noon the next day, including four re-counts, which reduced discrepancies. The party agents were meticulous. They said 140 votes missing and the ANC wanted to split the difference, but the DA party agents said 'no'".

The Inxuba Yethemba LM elections in 2021 were contested by ward level candidates from six political parties and independent candidates. Voter turnout was only 49%, which was marginally higher than the Eastern Cape average (47%), reflecting national trends of disillusionment with the ruling ANC. Proportionally, support for the ANC declined from 58.3% in 2016 to 52.4%, while DA support remained steady (39.2% to 39.8%) as did EFF support (2.5% to 2.3%) (Fig. 9.10).

The ANC received substantially fewer votes, as anticipated by the candidates to whom we had spoken. Local corruption replicates that at national level and was accordingly punished (albeit to a minimal extent) by voters. All three of the candidates whom we interviewed, were elected with strong majorities, suggesting that their personal popularity trumped any misgivings about the parties that they represented (Figs. 9.11 and 9.12).

In Ward 2, the eastern section of Lingelihle, the ANC's Mr. Monwabisi Rasmeni was elected as councillor with 1072 votes (93%), constituting a massive majority over the four other parties that contested the ward. The DA's Mr. Johannes Makatesi received 36 votes (3%), the EFF's Ms. Sheilla Rorani 32 votes (3%), the ATM's Ms. Nomthandiso Puwane 12 votes (1%), and the VF + 's Ms. Nicolene Hough, 2 votes (0.17%).

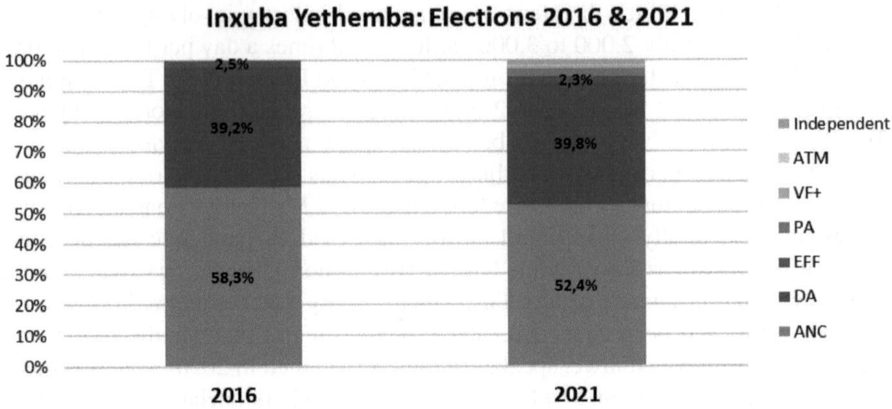

Fig. 9.10 Inxuba Yethemba local government results, 2016 and 2021. (Source: Derived from IEC, 2021)

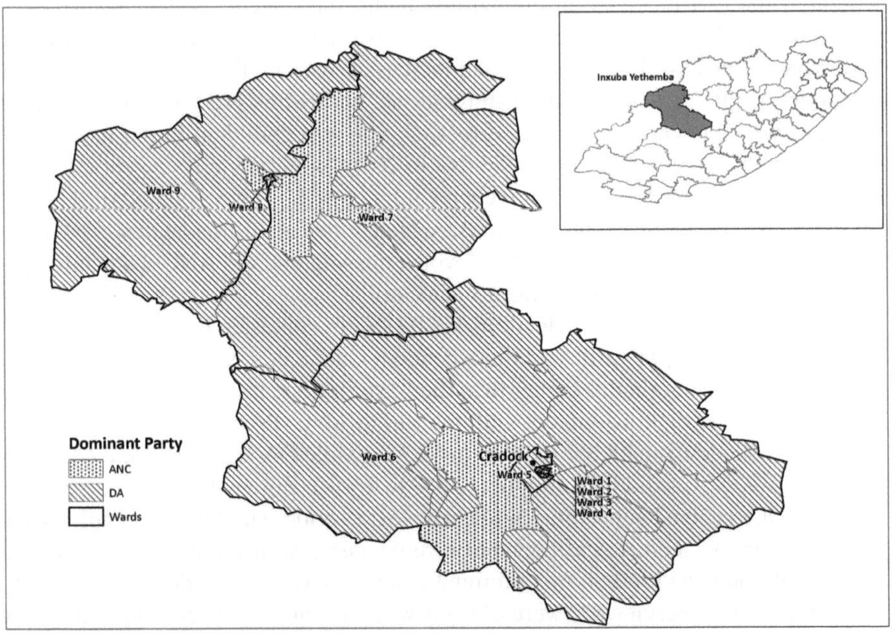

Fig. 9.11 Dominant party in each ward, Inxuba Yethemba, 2021. (Cartographer: Johann Fenske; data derived from IEC, 2021)

In Ward 3, the north-western section of Lingelihle, the ANC's Mr. Andile Dakuse was elected as councillor with 1181 votes (86%), ahead of DA Ms. Charmel Fleurs with 145 votes (11%), EFF Mr. Siphokazi Myosana 31 votes (2%), Independent candidate Ms. Dumisani Beja 13 votes (1%), VF+ Mr. Anton Mulder 2 votes (0.15%) and ATM Ms. Thandiwe Pikinini-Mtshemla 1 vote (0.07%).

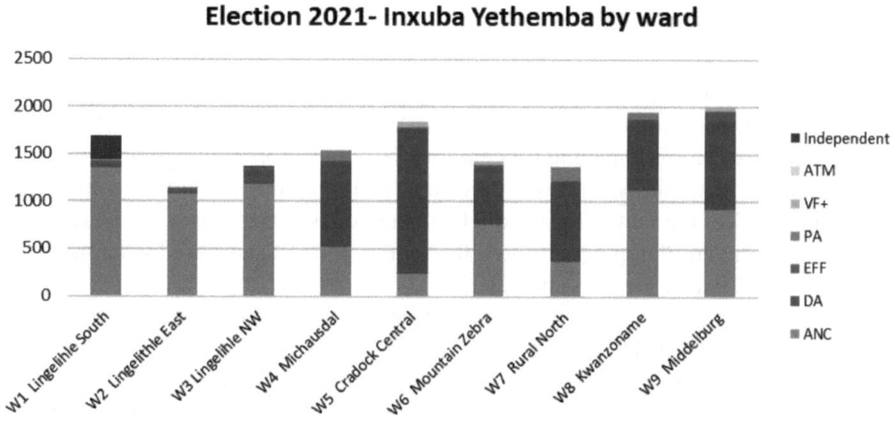

Fig. 9.12 Number of votes in each ward, Inxuba Yethemba, 2021. (Source: Derived from IEC, 2021)

In Ward 5, the central formal middle class section of Cradock, the DA's Mrs. Maria Hendrika Featherstonehaugh was elected as councillor with 1511 votes (81%), ahead of ANC Ms. Thembinkosi Nesi with 235 votes (13%), VF+ Ms. Nicolene Hough 50 votes (3%), PA Mr. Eugene Samuels 29 votes (2%), and EFF Mr. Vuzicelo Ysterman 21 votes (1%).

The eighteen elected LM council members comprised five wards councillors for the ANC (wards 1, 2 and 3 in Lingelihle; ward 6 rural Mountain Zebra area; and ward 8 Kwanzoname, adjacent to Middelburg) and four for the DA (ward 4 Michausdal; ward 5 Cradock central; ward 7 rural northern area; ward 9 Middelburg), supplemented by nine proportional representative councillors, five allocated to the ANC, three to the DA and one to the PA. The ANC PR candidate Ms. Noncedo Zonke was elected Mayor.

Subsequent to the elections, ongoing allegations of corruption lingered over the municipality. In February 2022, Council Acting Speaker Mr. Sibongile Masawe appointed Mr. Khathutshelo Mulaudzi as Chief Financial Officer "despite Council having resolved in January never to appoint [him] again". Additionally, the ANC called a special council meeting to appoint Mr. Mkhululi Mbebe as Municipal Manager, despite a case of stock theft pending against him. In the meantime, "the municipality's actual Speaker, Tembakazi Mbotya, [was] held captive and intimidated in her own home". The DA Eastern Cape Shadow MEC for CoGTA Ms. Vicky Knoetze reported that she had asked the COGTA MEC Ms. Xolile Nqatha more than once to intervene (Knoetze, 2022). Accordingly, Ms. Nqatha had written to the Council asking that the two officials should be suspended.

In June 2022, Councillor Rasmeni was still agonising over the Council's failure to suspend Acting MM Mkhululi Mbebe and CFO Mr. Lucky Mulaudzi. A brief Hawks investigation concluded that "there are no supporting documents, and without documents we are going nowhere, because nothing we can prove. It's going to be your words against their words."

9.7 Plettenberg Bay

Plettenberg Bay is the major urban centre of the Bitou Local Municipality in the Western Cape. It is situated along the coast, 215 km west of Gqeberha. The region was the home of Khoisan people for centuries prior to European colonial settlement. In 1771, a navigational beacon was erected to caution ships about the rocky coastline, and in 1776, the Dutch East India Company constructed barracks. Dutch governor, Joachim van Plettenberg, toured the region in 1778 and gave his name to the settlement (Afrikaner Geskiedenis, 2023).

The population has increased rapidly in recent years from 18,567 in 2001 to 49,162 in 2011 and 59,157 in 2016 to 65,879 in 2020 (StatsSA, 2011, 2016; Bitou, 2020, World Population Review, 2023). Residents comprise speakers of three main languages: 49% isiXhosa, 35% Afrikaans and 11% English (StatsSA, 2016). In 2011, almost half (49%) of the adult population was employed, and the mean household income was R29,400 (StatsSA, 2011). The population of the core urban settlement of Plettenberg Bay and suburban KwaNokuthula, New Horizons and Beacon Island Estate is 60% black Africans, 22% coloured, and 16% white (City Population, 2011).

More than four-fifths (82%, i.e. R541-million) of the municipal budget of R66-million was generated locally from payments made by residents for water, electricity, rates, licenses, fines, and from interest and investments (Municipalmoney, 2022). The balance (R123-million, i.e. 18%) comprised financial transfers from the 'Equitable Share' of taxes, and grants from national and provincial government. The Auditor-General gave a verdict on the municipality's finances as 'financially unqualified with findings' for 2021–11, after the previous 4 years' 'qualified with findings' (South Africa, 2022).

Bitou LM is divided into seven wards, five of which comprise the urban settlement of Plettenberg Bay. These are Ward 2 (central Plettenberg Bay, Piesangvallei, eastern New Horizons, Beacon Island Estate and Beacon Island), Ward 3 (central section of New Horizons), Ward 4 (western New Horizons), Ward 5 (northern KwaNokuthula) and Ward 6 (western and southern KwaNokuthula). Wards 1 and 7 comprise the rural hinterland of Plettenberg Bay to the north and west, respectively.

Ward 2 candidate for the DA, Mr. Dave Swart (Fig. 9.13) agreed to an interview and started by saying that he was "not a career politician". He and his wife had moved to the municipality in 2010 with the intention of getting out of the 'rat race' in Pretoria, where he was in 'sales'. They had operated various businesses in 'Plett' and he was elected as chairman of the Guest Accommodation Association. In 2017, when Councillor Mr. Wayne Craig resigned to pursue business interests, the DA approached him to stand in the by-election. As he was already dealing with role players, it seemed a logical "step". He was elected (86.33% of the 3543 votes cast with 66.25% voter turnout). Since then, he said, the "learning curve has been steep" but he is a "community-based person".

Mr. Swart asserted that for visitors, the town is neat and efficient as a result of private initiatives. The 'Keep it Clean' team, initiated by a local business owner

Fig. 9.13 Candidate posters in Bitou, 2021. (Photos: by author)

Alison Bryant (Stander, 2018) comprises weekly volunteer sessions to collect litter. This attracts funding from local businesses sufficient to employ 20 full-time workers. Similarly, crime statistics are "good" as a consequence of the efforts of the Plett Bay Crime Prevention Association, that monitors security and activates calls to contracted security companies in the town. Additionally, the widespread fire outbreaks in 2017 were largely brought under control by municipal firefighters with the assistance of local volunteers.

In stark contrast, poor performance amongst some key municipal professionals has tarnished the effectiveness of the municipality. Mr. Swart mentioned delays in the approval of house plans, and the transfer of properties owing to the appointment of unqualified officials. Additionally, there were instances of wastage and fraud in respect of municipal spending, frequent and sometimes irregular 'favour' senior appointments, inefficiencies in electricity and water meter readings, a salary bill amounting to 40% of municipal revenue, a sudden urgency in road maintenance the week before the elections. He said that in 2012 the DA inherited a R30-million budget deficit, but lost control of the council in 2016 when there was a R6-million positive balance. Amongst municipal employees "Good people get frustrated and leave or are forced into doing things that they are not comfortable with". He had been prevented from viewing some municipal documents but through making personal contacts and friendships, he was able to access information about suspicious tender awards and excessive payments to "dodgy suppliers". The DA had voted against the previous two annual budgets of around R750-million, but they were passed by the ANC-Active United Front (AUF) coalition of seven councillors, to the DA's six. There had not been a council meeting since June (interview conducted 28/10/21) when the Mayor was suspended.

Mr. Swart recognised that the ANC was tarnished by the high-profile corruption being exposed by the Zondo Commission, which discourages younger people from voting because they "don't feel that allegiance" that occurs amongst older generation ANC supporters. The DA was portrayed as racist by other parties although "90% of DA delegates in the Western Cape are not white" and he was "the only white candidate" for the DA in Bitou. His observation was that the EFF "appeals to people who have nothing by promising them everything". He nevertheless reported having a "good relationship with most" ANC councillors, and frequent interventions in wards other than his own Ward 2. He hoped that the DA would win a majority of wards because previously sharing power through a coalition had hampered progress. He predicted an improvement for the DA from the current six to seven or eight seats, based on local estimates of 48% DA support and 42% ANC support. He expected the DA to garner most votes in "well off" areas and increased support in the poorer "informal Xhosa areas". He expected continued AUF proportional representation in the council, improving on its fewer than 1000 votes in the previous election.

Tourism is a major economic component of the local economy, which generates income for local business from international and local visitors, and events such as the annual Matric Rave. He was optimistic that prospects would improve as the Covid-19 pandemic diminished and indicated that every Rand spent by visitors has a multiplier effect of 7 to 8.

A local alliance between the ANC and the AUF fell apart early in 2021 after a vote of no confidence. The AUF Mayor Mr. Peter Lobese resigned, and Mr. Bill Nel of the DA was elected mayor, with the support of Lobese. Nel was expected to run for mayor in the 2021 elections but withdrew for health reasons and Mr. Swart was nominated in his place.

Mirroring the national trends, the Bitou election results were disappointing for both the ANC and the DA (Fig. 9.14). Interest in the election was much higher than average, with a total of 25 political parties participating.

Eighteen of these fielded a candidate in one or more of the seven wards, and between them attracted 19,884 votes. A further seven parties fielded PR candidates only. A voter turnout of 64% was realised, as opposed to the much lower Western Cape mean of 49%.

The ANC lost one-third of its support, declining from 44% to 29% (Fig. 9.15), signifying a clear rejection of decades of poor governance at both national and local level. Support for the DA declined by a lesser margin from 46% to 40%, contrary to its hopes of gaining support at the expense of the ANC. The greatest change comprised dramatic growth in support for smaller parties, especially the AUF (5% to 9%), the Plett Democratic Congress (PDC) (0% to 7%), the PA (0% to 6%) and the Ikhwezi Political Movement (IPM) (0% to 3%).

The participation of Mr. Dave Swart (DA) attracted the overwhelming backing of voters in the affluent and predominantly white Ward 2 (Fig. 9.16). He received 3978 votes, comprising almost three times as many as any other candidate in any of the seven Bitou wards. Additionally, the DA achieved the largest proportion of votes in wards 1 and 7, and much smaller shares in poorer township and informal

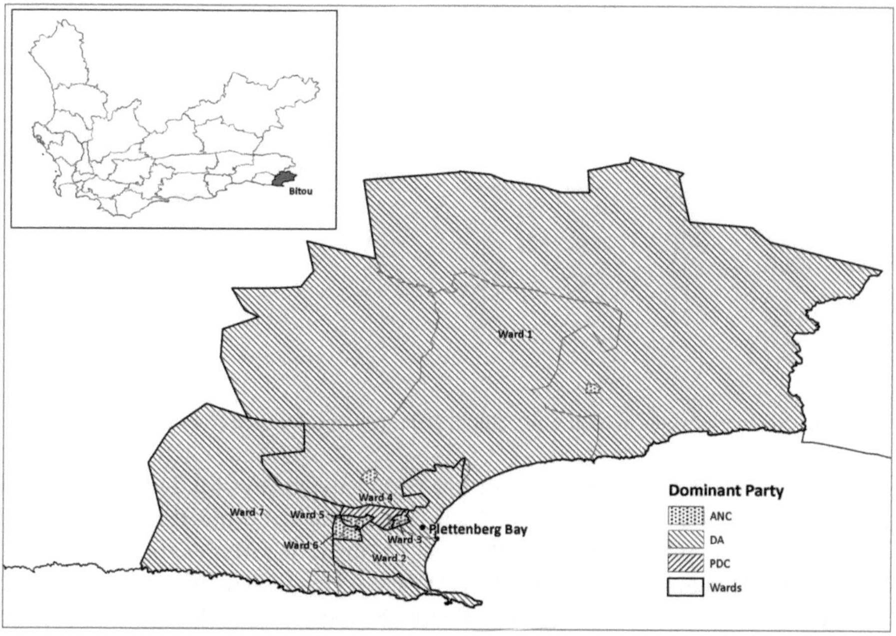

Fig. 9.14 Dominant party in each ward, Bitou, 2021. (Cartographer: Johann Fenske; data derived from IEC, 2021)

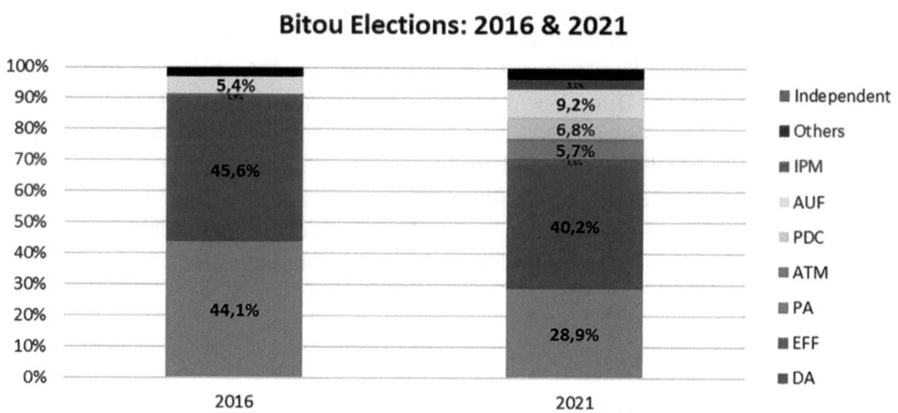

Fig. 9.15 Bitou local government results, 2016 and 2021. (Source: Derived from IEC, 2021)

settlements, namely wards 3, 4, 5 and 6. Conversely, the ANC attracted the largest proportions of votes in wards 3 (New Horizons east), 5 (KwaNokuthula north), and 6 (KwaNokuthula south). The PDC achieved the most votes in Ward 4 (New Horizons west), with a mainly coloured voter population, thereby defeating the incumbent DA councillor.

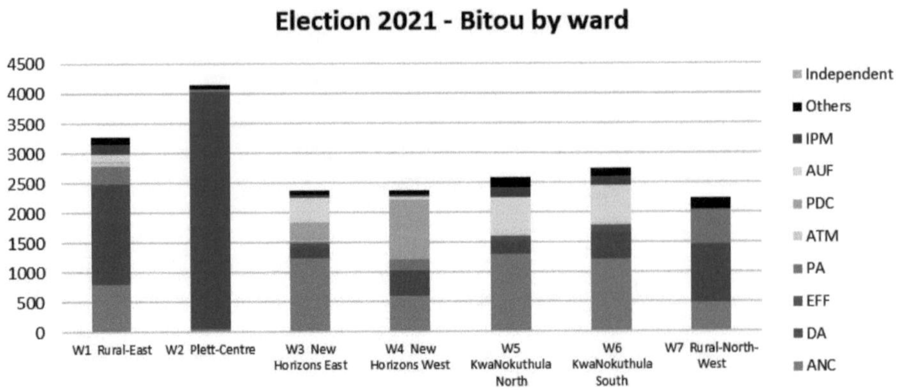

Fig. 9.16 Number of votes in each ward, Bitou, 2021. (Source: Derived from IEC, 2021)

In Bitou's Ward 2, with 3978 votes (96%) going to the DA's Mr. David Swart, not many were left for the ANC's Mr. Euan Wildeman (51), the PA's Ms. Andrea Thacker (33), the ACDP's Mr. John Ruiters (27), the IPM's Ms. Victoria Rowe (20), The Organic Humanity Movement (TOHM)'s Ms. Diana Fitzpatrick (18), the Cape Independence Party (CIP)'s Mr. Joseph Fourie (8), the EFF's Mr. Bandile Mlindazwe (5), and GOOD Party's Mr. Reynold Williams (1).

The thirteen elected LM council members thus comprised three ward councillors each for the DA (wards 1, 2, and 7) and the ANC (wards 3, 5 and 6) and one (ward 4) for the PDC. These were supplemented by six proportional representative councillors, allocated to the DA (2), AUF (1), PDC (1), PA (1) and IPM (1) respectively. Requiring seven councillors to constitute a majority, the DA, AUF and PDC opted to work together. At the Bitou Local Municipal Council meeting of 18th November 2021, Councillor Dave Swart (DA) was elected as Executive Mayor and Councillor 'Punki' Busakwe (AUF) was elected as Deputy Mayor (Bitou, 2021). Mayor Swart, in June 2022 told me that "our Coalition partners are on the same page with our endeavours to restore proper governance to Bitou and rebuild our finances to a position where, in future, we can properly plan and align our activities to service delivery issues".

9.8 Discussion and Conclusions

The two dominant political parties in South Africa each lost substantial support in the November 2021 local government elections. This chapter illustrates the dynamics and configurations of this outcome in four local municipalities: Karoo Hoogland (Northern Cape), Kopanong (Free State), Inxuba Yethemba (Eastern Cape) and Bitou (Western Cape). The research entailed interviews with local government election candidates in the four municipalities. They were unanimous in identifying corruption and low levels of delivery of municipal services as the prevailing issues in need of attention. They and other local activists had often intervened in the absence of municipal action, to mitigate inadequacies in the delivery of services. In all four municipalities, the shares of votes received by the ANC and DA declined, although individual candidates from these parties who had proven records of competence and dedication fared better. Smaller local parties perceived to be more aware of local needs, received increased proportions of the vote.

Auditor-General Ms. Tsakani Maluleke asserts in her 2021–22 report (South Africa, 2022) that "courageous, ethical, accountable and citizen-centric leadership is needed to turn the tide in local government". In the absence thereof, our research suggests the need for a further and perhaps more realistic and more achievable requirement to the AG's criteria: namely, sustainable voluntary initiative and intervention. One of the keys to achieving this is the social capital of the 'creative class' (Ingle, 2009), comprising people who make a living from creative pursuits, including artists, designers and knowledge-based professionals. A core group of self-motivated entrepreneurial types are conducive to economic multipliers across a local economy. Another key is the emergence of community activists with the capacity and determination to assist with the technical tasks which a municipality is failing to do, such as garbage disposal, water and electrical service maintenance. Configurations of collaborative community-based efforts, as identified in the four towns in this chapter, serve to monitor, supplement and complement the usually grossly inadequate efforts of the professional municipal workforce.

Elections analyst Wayne Sussman (2021) identified the strongly localised self-help orientated response to the elections across the country. He observed that "In a record number of smaller municipalities, voters decided to walk away from the ANC, DA and even the EFF, and support local parties. Voting for a party that is stationed in your far-flung municipality is now more appealing than voting for a deployee from Luthuli House or the DA's national office". He commented further that "[T]hese parties promised local solutions to local challenges, rather than a half-baked cookie-cutter approach to endemic problems. They usually fielded candidates who had defected from one of the major parties or people who had a strong standing in the local community". In the four towns spotlighted in this chapter, local dynamics played a prominent role in disrupting the long-term hegemony of the ANC, and to a lesser extent, the DA, in their traditional vote-catching moulds.

Sutherland (Karoo Hoogland LM) is now governed by an alliance of three minority parties. The DA + PA + VF+ (namely, 3 + 2 + 1 = 6), achieved the six seats required to wrest control of the Council from the ANC, which won 5 seats, with voter turnout holding steady at 62%. The ANC had previously won a majority in 2016 of 4 to 3 (DA 2 and COPE 1). In Philippolis (Kopanong LM), the ANC's previous massive 11 seat majority over the DA's 2 and EFF's 1, was slightly diluted to ANC 11, DA 3, EFF 2 and VF+ 1, with several new faces in the ANC caucus. However, a large decline in voter turnout from 62% in 2016 to 49% in 2021, indicated that many voters preferred to abstain rather than to cast their votes for alternative parties.

In Cradock (Inxuba Yethemba LM), the ANC:DA seat ratio of 11:7 changed to ANC10: DA7: PA1. Voter turnout declined substantially from 57% to 47%, also illustrating that the preferred means of punishing the ANC was to abstain from voting. Plettenberg Bay (Bitou LM) saw a decline in the support given to both the ANC and the DA, voter turnout dropping marginally from 68% to 64%, but with quite dramatic increases in support for smaller parties. Two of the latter (AUF and PDC) aligned with the DA to facilitate a majority 7–6 over the ANC (4) and any potential alliance that it might have contemplated with the PA (1) and IPM (1). This enabled the DA-PDC-AUF to assume control of the council.

Acknowledgement The enthusiastic support of Susan Rule during the cross-Karoo data gathering trip was highly valued. The expert cartographic skills of Johann Fenske are evident in the maps.

References

Afrikaner Geskiedenis. (2023, March 2023). *Joachim van Plettenberg (1739–1793)*. Joachim van Plettenberg (1739-1793) – Afrikanergeskiedenis

Alexander, P., Runciman, C., Ngwane, T., Moloto, B., Mokgele, K., & Van Staden, N. (2018). Frequency and turmoil: South Africa's community protests 2005–2017. *South African Crime Quarterly, 63*, 2–42.

Beinart, W. (2008). *The rise of conservation in South Africa: Settlers, livestock, and the environment 1770-1950*. Oxford University Press.

Binza, S. M. (2005). The evolution of south African local governments: The politics of memory. *Journal of Contemporary History, 30*(2), 69–87.

Binza, M. S. (2010). Developmental local government in the 21st century: Implementing local economic development as a new mandate. *Administratio Publica, 18*(4), 241–260.

Bitou. (2020). *Socio-economic profile*. Bitou Municipality 2020. Western Cape Government. https://www.westerncape.gov.za/provincial-treasury/sites/provincialtreasury.westerncape.gov.za/files/atoms/files/SEP-LG%202020%20-%20WC047%20Bitou%20Municipality.pdf

Bitou. (2021, November 2021). *Council meeting*. YouTube. https://www.youtube.com/watch?v=vGVrbsMQM-g.

Blackman, M., & Dall, N. (2022). *Spoilt Ballots: The elections that shaped South Africa, from Shaka to Cyril*. Penguin.

Booysen, S. (2004). Ten years of democracy and the state of opposition politics in South Africa. *Election Update, 2004*(8), 9–12.

Booysen, S. (2015). Election 2014 and the ANC's duet of dominance and decline. *Journal of African Elections, 14*(1), 7–34.

Butler, A. (2014). The ANC's campaign in 2014. In C. Schulz-Herzenberg & R. Southall (Eds.), *Election 2014 – The campaigns, results and future prospects* (pp. 42–56). Jacana & Konrad Adenauer Foundation.

Calland, R. (2013). *The Zuma years: South Africa's changing face of power*. Random House Struik.

Christopher, A. J. (1996). Regional patterns in South Africa's post-apartheid election in 1994. *Environment and Planning C Government and Policy, 14*, 55–69.

City Population. (2011). *citypopulation.de*.

Dhawraj, R., Du Plessis, D., & Du Plessis, C. (2021). South Africa's 2016 municipal elections how the ANC and DA leveraged twitter to capture the urban vote. *Journal of African Elections, 20*(1), 1–22.

Donaldson, R. (Ed.). (2017). *Small town tourism in South Africa*. Springer.

Donaldson, R., & Marais, L. (Eds.). (2012). *Small town geographies in Africa: Experiences from South Africa and elsewhere*. Nova Science Publishers.

Fox, R., & Lemon, A. (2000). Consolidating South Africa's new democracy: Geographical dimensions of party support in the 1999 election. *Tijdschrift voor Economische en Sociale Geografie, 91*, 347–360.

Friedman, S. (1999). Who we are: Voter participation, rationality and the 1999 election. *Politikon, 26*, 213–223.

Friedman, S. (2004). Why we vote: The issue of identity. *Election Synopsis 1*, 2–4. Human Sciences Research Council.

Gvaramadze, V. V., Kniazev, A. Y., Maryeva, O. V., & Berdnikov, L. N. (2018). Optical spectroscopy of the blue supergiant Sk−69° 279 and its circumstellar shell with SALT. *Monthly Notices of the Royal Astronomical Society, 474*(1), 1412–1425.

Hoogendoorn, G., & Nel, E. (2012). Exploring small town development dynamics in rural South Africa's post-productivist landscapes. In R. Donaldson & L. Marais (Eds.), *Small town geographies in Africa* (pp. 21–34). Nova Science Publishers.

Independent Electoral Commission of South Africa (IEC). (2021). https://results.elections.org.za/dashboards/lge/

Ingle, M. (2009). A 'creative class' in South Africa's arid Karoo region. *Urban Forum, 21*(4).

Ingle, M. (2010). Making the Most of 'nothing': Astro-tourism, the Sublime, and the Karoo as a 'space destination'. *Transformation, 74*, 87–111.

Karoo-Hoogland Local Municipality. (2022). *Draft 5th Generation Integrated Development Plan*.

Khosa, M. M., & Muthien, Y. G. (1998). *Regionalism in the new South Africa*. Ashgate.

Knoetze, V. (2022, March 1). *Inxuba Yethemba Council Defiance Continues*. My Gqeberha. Inxuba Yethemba council defiance continues (mygqeberha.com).

Kopanong Local Municipality. (2011). *Kopanong Local Municipality Annual Financial Statements for the year ended 30 June, 2011*.

Kotze, J. S., & Bohler-Muller, N. (2020). Introduction: Quo Vadis South Africa? Reflections on the 2019 South African General elections. In J. S. Kotze & N. Bohler-Muller (Eds.), *Reflections on the 2019 South African General elections: Quo Vadis?* Routledge.

Lemon, A. (2009). The general election in South Africa, April 2009. Notes on recent elections. *Electoral Studies, 28*, 642–673.

Lodge, T. (1999). *Consolidating democracy: South Africa's second popular election*. Witwatersrand University Press.

Mabin, A. S. (1984). *The making of colonial capitalism: Intensification and expansion in the economic geography of the Cape Colony, South Africa, 1854-1899*. PhD thesis, Department of Geography, Simon Fraser University. https://summit.sfu.ca/_flysystem/fedora/sfu_migrate/5870/b1529402x.pdf

Mabin, A. (2021). History and hope in Cradock, Eastern Cape. *Transformation, 106*, 35–67.

Maharaj, B. (2009). Migrants and urban rights: Politics of xenophobia in South African cities. *L'Espace Politique, 8*, 1–15.

Maharaj, B., & Ramutsindela, M. (2002). Introduction: Post-apartheid political dispensation – New or old geographies? *GeoJournal, 57*(1), 1–2.

Mattes, R., & Piombo, J. (2001). Opposition parties and the voters in South Africa's general election of 1999. *Democratization, 8*, 101–128.

Mattes, R., & Southall, R. (2004). Popular attitudes towards the South African electoral system. *Democratization, 11*(1), 52–77.

McLaughlin, E. S. (2007). Beyond the racial census. *Comparative Political Studies, 40*, 435–456.

Melber, H. (2014). South Africa's elections 2014: More than more of the same? *Review of African Political Economy, 41*, 645–651.

Montsho, M. (2016, July 28). *EFF blasts north west speaker for 'electioneering'*. Independent Online (IOL).

Municipalmoney. (2022). https://municipalmoney.gov.za/profiles

Muthien, Y. (1999). *Democracy South Africa: Evaluating the 1999 election*. Human Sciences Research Council.

Myburgh, P. L. (2019). *Gangster state: Unravelling ace Magashule's web of capture*. Penguin.

Nel, A. C. (2016). *Speech by the deputy minister for cooperative governance and traditional affairs*. Presentation at HSRC Seminar.
Netswera, M. M. (2012). Local government elections trends: 2000–2011. *Politeia, 31*(1), 74–89.
Ngoma, A. (2014). Black professionals and the ANC in the 2014 election: Loosening ties? In C. Schulz-Herzenberg & R. Southall (Eds.), *Election 2014 – The campaigns, results and future prospects* (pp. 155–168). Jacana & Konrad Adenauer Foundation.
Noble, M., & Wright, G. (2013). Using indicators of multiple deprivation to demonstrate the spatial legacy of apartheid in South Africa. *Social Indicators Research, 112*, 187–201.
Olivier, J. L., & Rule, S. (1999). Free and fair: Voters' evaluation of the 1999 election. *Politikon, 26*, 225–234.
Paret, M., & Bekker, M. (2021). *Voter fluidity in the 2021 Local Government Elections: Evidence from five metropolitan municipalities*. University of Johannesburg.
Pauw, J. (2017). *The President's keepers: Those keeping Zuma in power and out of prison*. Tafelberg.
Prinsloo, J. (2014). Examining the relationship between the news media and the ANC: A case study of Nkandla coverage. *Ecquid Novi: African Journalism Studies, 35*(2), 23–39.
Ramutsindela, M. F. (1998). Afrikaner nationalism, electioneering and the politics of a Volkstaat. *Politics, 18*, 179–188.
Ramutsindela, M. F. (2001). Down the post-colonial road: Reconstructing the post-apartheid state in South Africa. *Political Geography, 20*, 57–84.
Reynolds, A. (1999). *Election '99 South Africa: From Mandela to Mbeki*. David Philip.
Roberts, B., Struwig, J., Gordon, S. L., & Davids, Y. D. (2020). The unconvinced vote: The nature and determinants of voting intentions and the changing character of South African electoral politics. In J. Steyn Kotze & N. Bohler-Muller (Eds.), *Reflections on the 2019 South African General Elections: Quo Vadis?* Routledge.
Rule, S. (2000). *Electoral territoriality in Southern Africa*. Ashgate.
Rule, S. (2018). Geography and voting: The growth of urban opposition in South Africa two decades after democratisation. *South African Geographical Journal, 100*(2), 141–161.
Runciman, C., Alexander, P., Rampedi, M., Moloto, B., Maruping, B., Khumalo, E., & Sibanda, S. (2016). *Counting police-recorded protests: Based on South African police service data*. Social Change Research Unit, University of Johannesburg.
Russon, R. D. (2011). Ten years of democratic local government elections in South Africa: Is the tide turning? *Journal of African Elections, 10*(1), 74–98.
Sadie, Y., & Patel, L. (2020). Zuma versus Ramaphosa: Factors influencing party choice of south Africans in the run-up to the 2019 elections. *Journal of African Elections, 19*(1), 1–26.
Schulz-Herzenberg, C. (2013). The implications of social context partisan homogeneity for voting behaviour: Survey evidence from South Africa. *International Journal of Social, Human Science and Engineering, 7*, 983–999.
Seethal, C. (2012). South Africa's local government elections of 2011. *South African Geographical Journal, 94*, 9–21.
South Africa. (2019). *Report of the auditor-general to the Free State legislature and the council on the Kopanong local municipality*. Auditor-General, 31 January 2019.
South Africa. (2000). *Local Government Municipal Systems Act 32 of 2000*. https://www.gov.za/documents/localgovernment-municipal-systems-act
South Africa Auditor General. (2022). *MFMA local government audit outcomes. Annexure 3: Auditees' audit opinions over the past five years*. Auditor-General of the Republic of South Africa. https://www.agsa.co.za/Reporting/MFMAReports/MFMA2020-2021.aspx
Southall, R. (2014). The black middle class and democracy in South Africa. *Journal of Modern African Studies, 52*, 647–670.
Southall, R., & Schulz-Herzenberg, C. (2014). The party system and political prospects in the wake of election 2014. In C. Schulz-Herzenberg & R. Southall (Eds.), *Election 2014 – The campaigns, results and future prospects* (pp. 228–240). Jacana & Konrad Adenauer Foundation.

Southern African Large Telescope. (2021). Southern African Large Telescope. *Annual Report.* Cape Town/ Sutherland. https://www.salt.ac.za/wp-content/uploads/2022/04/SALT_AR_2021

Spence, J. E. (Ed.). (1999). *After Mandela: The 1999 South African elections.* The Royal Institute of International Affairs.

Stander, Y. (2018, August 31). *Plett woman keeps it clean.* Knysna-Plett Herald.

StatsSA. (2016). *Community survey 2016.* Interactive graphs. www.statssa.gov.za

StatsSA. (2011). *Pop census 2011.* www.statssa.gov.za

Sussman, W. (2021). *Locally focused parties the biggest winners of the municipal elections.* WAYNE SUSSMAN: Locally focused parties the biggest winners of the municipal elections (businesslive.co.za).

Visagie, J. (2013). *Econ3x3.* Retrieved from http://www.econ3x3.org/article/who-are-middle-classsouth-africa-does-it-matter-policy

World Population Review. (2023). *Plettenberg bay population 2023.* https://worldpopulationreview.com/worldcities/plettenberg-bay-population

Worrall, D. (1971). *South Africa. Government and politics.* LJ van Schaik.

Zondo, R. M. M. (2022). *Judicial Commission of Inquiry into allegations of state capture, corruption and fraud in the public sector including organs of state report: Part VI Vol. 4: Summary of recommendations chairperson: Justice R.M.M Zondo Chief Justice.* https://static.pmg.org.za/State_Capture_Commission_Report_Part_VI_Vol_IV.pdf

Chapter 10
The Historical Development of South African Small Towns as Spa Resorts

Christian M. Rogerson and Jayne M. Rogerson

10.1 Introduction

Tourism research on South African small towns must be understood as falling within the neglected 'other half' of urban tourism scholarship as relates to the Global South (Rogerson & Rogerson, 2021a). Until the mid-2000s there existed only a minimal focus on small town tourism as a research issue in South Africa (Donaldson, 2007, 2018). Since that time, however, there has been a burst of scholarship and fresh insights emerging on various facets of tourism development in small town South Africa. As argued by Visser (2016) and Rogerson and Visser (2020) although contributions have been made by several disciplines the major share of the new research has been generated by local tourism geographers. In rich overviews of the state of the art literature on small town tourism in South Africa, Donaldson (2018, 2021) identifies the strength of writings especially on second homes, festivals and events, as well as nature-based tourism and rural dynamics.

Arguably, given the economic and employment challenges faced by small towns, the largest amount of research surrounds the vital importance of tourism as a potential driver for local economic development (Houghton et al., 2013; Donaldson, 2018; Kontsiwe & Visser, 2019; Rogerson & Rogerson, 2019, 2020a; Dlomo & Rogerson, 2021). Related themes have focussed on questions of inclusive development (Butler & Rogerson, 2016; Rogerson & Rogerson, 2020b), the imperative for maximising local municipal assets (Rogerson, 2020; Rogerson & Rogerson, 2021b), local management challenges (Wessels & Tseane-Gumbi, 2022), tourism-led gentrification processes (Donaldson, 2009, 2018), and the potential of leveraging various forms of niche tourism including adventure tourism (McKay, 2013, 2020), agritourism (Rogerson & Rogerson, 2014), astro-tourism (Jacobs et al., 2020),

C. M. Rogerson · J. M. Rogerson (✉)
School of Tourism & Hospitality, University of Johannesburg, Johannesburg, South Africa
e-mail: jayner@uj.ac.za

© The Author(s), under exclusive license to Springer Nature Switzerland AG 2023
R. Donaldson (ed.), *Socio-Spatial Small Town Dynamics in South Africa*, GeoJournal Library, https://doi.org/10.1007/978-3-031-37142-4_10

creative tourism (Drummond, 2021), gastronomy (Rogerson & Rogerson, 2021c), heritage (Drummond et al., 2021), slow tourism (Donaldson, 2018, 2021), and wine (Booyens, 2020). Route tourism promotion has been a long-established theme in investigations of small town tourism in South Africa (Lourens, 2007; Rogerson, 2007; Donaldson, 2018). More recently, the ramifications of the COVID-19 pandemic for the tourism prospects of small towns (Rogerson & Rogerson, 2022a, b; Rogerson et al., 2021a, b) as well as the adaptive responses made by local firms to the COVID-environment (Giddy & Rogerson, 2021; Rogerson, 2021; Giddy et al., 2022) have been pinpointed in local research.

Overall, it is apparent that the majority of these studies engaging with small town tourism in South Africa concentrate on contemporary policy-relevant issues. This emphasis is unsurprising given the directions of mainstream tourism scholarship (Visser, 2016; Rogerson & Visser, 2020). The bias is accounted for by the critical role of tourism for policy development at the local level and especially its perceived potential as a driver for small town local development prospects in South Africa (Rogerson, 2014; Kontsiwe & Visser, 2019). Only a handful of research studies explore the historical dimensions of tourism in small town South Africa. One of the earliest contributions was by Davidson (2000) who examined the nineteenth century aspirations of Grahamstown to become 'the Saratoga of South Africa', a place for health and relaxation. More recent research has been produced about the development of accommodation services (Rogerson, 2019), the early evolution and role of caravan parks (Rogerson & Rogerson, 2021d), the apartheid rise and fall of adventure tourism activities such as skydiving (Drummond et al., 2022), the nineteenth century appearance of several small towns as health resorts (Rogerson & Rogerson, 2021e) and, the making of small towns as coastal leisure destinations (Rogerson & Rogerson, 2020c).

Building upon these foundations this chapter is a further contribution to address the 'present-mindedness' of mainstream small town tourism research investigations in South Africa. The merits of undertaking such historical research are stressed by Walton (2009: 115) as all tourism research "needs a sense of historical awareness" because "the present cannot be understood without reference to what has gone before". The aim is to examine the historical trajectory of small towns as destinations for spa tourism. The chapter adopts an historical approach and draws upon archival sources mainly from the collections at the National Library depot in Cape Town. The time period under investigation begins in the late nineteenth century and concludes in 2000. Three sections of material are presented. In the following section, a context is provided in terms of reviewing international literature which has explored the historical making and evolution of spa resorts. Turning to South Africa the subsequent sections discuss factors influencing the role of mineral springs in the historical tourism development of the country's small towns before moving to a local case study of Caledon as one spa destination in the Western Cape.

10.2 Spa Resorts – An International Context

The idea of the 'spa' has a long history and tradition (Walton, 2012; van Wyk, 2013). It is argued that the spa resort can be viewed as a distinctive category of 'leisure resort' as compared, for example, with the coastal resort, the colonial hill station or the resort where the major assets are natural scenery, culture or heritage (Walton, 2014). Frost and Laing (2016) provide a history of 'spa tourism' dating back to ancient times when many people travelled to places boasting hot springs several of which were based on sacred sites. In the historical usage of the term a "spa resort" is a mineral springs resort at which visitors 'take the waters' either by bathing or drinking them while enjoying leisure, entertainment and sociability (Walton, 2012). In Britain the term 'spa', derived from the name of the pioneer international resort of this type which is now in Belgium, was applied to all non-coastal bathing resorts as well as to medicinal springs located in coastal settings such as the Yorkshire resort of Scarborough (Brodie, 2012). Historically, therefore the term spa includes cold baths – fashionable in England from the late seventeenth century – as well as warm or hot mineral springs (Walton, 2011). As pointed out by Walton (2012: 2) the modern terminology to describe a hotel that offers specialized bathing and massage services with no association to local mineral springs represents a privatisation of the concept applied to individual businesses rather than to localities or municipalities. In this chapter the older usage definition of the term spa is applied rather than the modern-day terminology.

Historically, spas have been founded in small town localities with local natural assets and "usually seen as symbols of local identity" (Walton, 2012: 2). According to Walton (2011: 138) the spa resort is essentially "a European phenomenon which has its counterparts in other parts of the world". Hembry (1990) maintains that beginning in the sixteenth century with the reign of Queen Elizabeth I members of the English nobility and landed gentry established the practice of taking relaxation at the country's inland spas. Walton (2014) points out that spa resorts were favoured destinations in Europe for affluent seekers of health and comfort in opulent surroundings from the seventeenth century. Brodie (2012) stresses that in the early eighteenth century personal health was of major concern and that in order to address chronic medical conditions and psychological disorders the wealthy and elite classes headed to spas where they received treatment including often with the recommendation for cold water bathing on medical grounds. In addition, according to Borsay (2012) another major reason that "spas grew so rapidly from the late seventeenth century was that they were not only centres of medicinal treatment but were also at the forefront of the provision and development of a new culture of fashionable leisure and tourism that acquired a powerful hold on the minds of the elite and those who aspired to join them" (Borsay, 2012: 156.). Rotherham (2014) contends that the British spa came into its own during the Georgian period with thousands of travellers seeking to partake of the waters of Bath, Cheltenham and Tunbridge Wells. The town of Bath, which was styled by Johnson (2013: 25) as "the Queen of the spas" was firmly situated at the top of the hierarchy of England's leading watering places.

Arguably, the economic prosperity of many pre-existing small towns in England was boosted and their populations expanded under the influence of the asset of local waters. Borsay (2012: 159) states that the spa resorts "oozed urbanity" and "were the places to which the fashionable gentleman, his wife and eligible daughters could migrate each season to discover the latest cultural trends, find excitement and sociability, and escape the isolation, boredom and backwardness of the countryside". From the eighteenth century there was considerable interest in 'taking the waters' at thermal or mineral baths most especially in Europe for their therapeutic qualities (Frost & Laing, 2016). This said, Brodie (2012) asserts that, notwithstanding an appetite for visits to spas in search of treatments for physical, mental and imaginary conditions, for the new leisured class of England "in modern terms most were probably simply seeking to improve their well-being" (Brodie, 2012: 125). As Walton (2011) shows during the first half of the nineteenth century many spa resort towns in England lost their economic dynamism because domestic customers increasingly sought out health and entertainment at spa resorts in France, Germany, Spain and even in more distant destinations. During the second half of the nineteenth century another contributory factor of the general decline in the comparative significance of English spas was the expanding popularity of seaside resort destinations the competitiveness of which began to eclipse the spas (Walton, 1983, 2011).

An international lens discloses the observation that mineral springs are to be found all over the world with their localities in various stages of development from 'discovery' to situations of rise and decline in a parallel to Butler's classic tourist area life cycle model of resort development (Walton, 2014). The development pathways of individual spa resorts in several countries have attracted a growing international scholarship (Walton, 2012). Smith (2008) traces the development and regeneration of the town of Spa in Belgium, one of the oldest and most famous thermal resorts in the world. Alonso-Álvarez (2012) tracks 'the value of water' as an asset for the origins and expansion of thermal tourism in Spain. Gordon (2012) records the establishment of Vichy as an international spa tourism destination which was propelled by the promotion of hydrotherapy by business and medical interests. The spring of Vichy has been styled as "the crown jewel of the French Spas" (van Wyk, 2013: 36). Beyond Europe there are several other investigations concerning the history of mineral spas for example, in Australia (White, 2012), the United States (Cross, 2012), Canada, New Zealand (Walton, 2014) and Japan (McMorran, 2008; Frost & Laing, 2016). Overall the largest literature relates to the historiography of spa resorts in the case of Britain, including their rise and fall (Hembry, 1990; Walton, 2011; Borsay, 2012; Brodie, 2012; Walton, 2012; Rotherham, 2014; Walton, 2014). In particular, the evolutionary trajectory of the town of Bath as a pioneer 'spa resort' has attracted much attention in the abundant scholarship about English spa resorts (Hembry, 1990; Walton, 2012; van Wyk, 2013).

10.3 South Africa – Mineral Springs and the Evolution of Small Towns as Spa Destinations

In his global historical survey of spa resorts and spas in history Walton (2012: 4) draws attention to "the plentiful supply of mineral spring spa resorts in South Africa". A decade ago the works of Boekstein (2012, 2014a, b, c) and Boekstein and Spencer (2013) were the first to focus directly on the importance of thermal springs for tourism development in small town South Africa. Another valuable contribution is that authored by van Wyk (2013) on the social history of three thermal mineral resorts in the Western Cape and their influence on the development of the modern health and wellness sector in South Africa. It has been estimated South Africa has approximately 80 thermal springs albeit only one-third of these have been used for tourism development (Boekstein & Spencer, 2013). The local economic base of several small towns has been expanded by seeking to utilise the asset of thermal springs. Among many small towns where mineral springs have contributed to local tourism development are Aliwal North, Barrydale, Calitzdorp, Cradock, Florisbad, Machadodorp (eNotokozweni), Malmesbury, Montagu, Rawsonville, Tshipise and Worcester. The most mature for tourism development are those located at Bela-Bela (former Warmbaths) in Limpopo, Badplaas in Mpumalanga, and at Caledon in the Western Cape (Boekstein, 2014a). It is observed, however, that the character of resort development linked to these centres varies quite considerably in terms of the size as well as the range of facilities on offer (Boekstein, 2012).

As a broad historical canvas to small town spa tourism Boekstein (2012) records that at the time of Dutch settlement in South Africa 'the taking of the waters' had become a well-established tradition among the upper classes of Europe. Accordingly, the Dutch settlers "arrived in South Africa with a well-developed spa culture, and it was not long before they discovered thermal springs in the vicinity of what is today the city of Cape Town" (Boekstein, 2012: 31). This said, it is acknowledged that most of South Africa's thermal springs have been used as places of healing by indigenous communities, even prior to the seventeenth century Dutch colonization (Rindl, 1936; Boekstein & Spencer, 2013; van Wyk, 2013). With British occupation and an expanded cultural influence van Wyk (2013: 52) asserts "it is therefore not a surprise to learn that the thermal water culture had travelled" to the Cape Colony. Moreover, "for the new colonisers the natural environment of the country was one to be explored and turned into a health related asset" (van Wyk, 2013: 52).

During the late nineteenth century South Africa emerged as a health destination because of the belief which was widespread in Britain and Europe that travel to places with dry air and high altitudes was beneficial to tuberculosis sufferers (Symes, 1889). As van Wyk (2013: 53) maintains "South Africa's status as a health resort was achieved, because it was the healing of this dreadful and common ailment that South Africa's climate was famous for". The 'discovery' that air and climate might assist in curing the condition of consumptives was especially significant in the context of imperial Britain which occupied South Africa during the nineteenth century. Several factors functioned to build the Cape Colony as an attractive health destination by the mid-nineteenth century (Bell, 1993; Rogerson & Rogerson, 2021e). These included a number of promotional initiatives in the form of guidebooks that gave information on the health opportunities of South Africa, and not least also for its several mineral spring resorts (Burton, 1907). At the time of the formation of the Union of South Africa in 1910, however, commercial tourism as an organized industry was still a relatively new phenomenon (Norval, 1936). A critical factor in the early development of tourism in South Africa was railway network expansion which improved access for both local and overseas visitors to the country's mineral springs. One example is the small town of Montagu where the railways facilitated access by motor coach from Ashton station on the New Cape Central Railway. The mineral springs there were likened to "Gastein" in Europe and described in one guide book as "more Radio-Active than the waters at Karlsbad and Baden-Baden" (Union Publishing Agency, 1923: 112).

According to Pirie (2011) the energetic promotion of international tourism to South Africa commenced when the newly-established, unitary government-owned South African Railways and Harbours (SAR&H) opened a publicity and travel department. Indeed, the associated promotional advertising and pamphlets that were produced by the SAR & H organisation from 1910 were foundations for the growth of both domestic and international tourism (Foster, 2003). Beginning in the 1920s and continuing for (at least) 25 years into the apartheid period SAR & H produced a series of publications which showcased the medicinal springs assets of South Africa with a view to their tourism promotion possibilities especially for overseas visitors (South African Railways and Harbours, 1928, 1930; Rindl, 1936; Carlyle-Gall, 1937; Kent, 1952). The first booklet appearing in 1928 focused on giving prominence to those small towns possessed both with medical springs and "suitable accommodation for those who are anxious to benefit by the curative properties of the waters" (South African Railways and Harbours, 1928: 3). Among the several profiled localities were included material on Aliwal North, Caledon, Florisbad, Machadodorp, Montagu, Warmbaths, Winburg and Worcester. The pamphlet emphasized greatly the asset of South Africa's climate and the potential combination of sunshine and medicinal waters in effecting cures. In addition, the benefits of using SAR & H were highlighted as follows: "the ample railway communications offer the best of facilities for comfortable travel over long distances" such that "when the visitor tires of the seeming monotony of the 'cure' it is an easy matter to find pleasurable diversion" (South African Railways and Harbours, 1928: 3).

In a 1930 publication devoted specifically to the 'wonder pools' of Aliwal North emphasis is given to the transformation and 'new lease of life' of the mineral springs after the local municipality relinquished control of the springs to private sector owners (South African Railways and Harbours, 1930: 13). It was noted that prior to this hand-over the Aliwal North springs had recorded a "rather lukewarm popularity" which resulted in "the far from meteoric career of this town" (South African Railways and Harbours, 1930: 9). In the local promotional material Aliwal North was styled as a 'welcome-stranger' town which was markedly different to European spa resorts:

Aliwal North cannot, by any stretch of the imagination, be likened to such resorts of the *beau monde* as Cannes, Carlsbad or Biarritz; in fact, it is doubtful whether the people of Aliwal would feel at all flattered were the comparisons made. The hectic life in evidence at the Continental spas does not manifest itself at this small South African town with its one business street. But in Aliwal North, unpretentious and without social, theatrical or architectural orientation, the visitor is assured of one thing – a welcome (South African Railways and Harbours, 1930: 23).

By the mid-1930s the targeting of promotional literature at the potential international tourist was once again strongly in evidence. The merits of South Africa's local medicinal springs to overseas visitors were viewed as deserving of greater attention. It was described that: "Here and there throughout the land warm springs, possessing health-giving properties, bubble up from the bowels of the earth, and those in search of 'a cure' can be assured of finding several sanatoria where they will find the hours pass in a round of pleasure, the while their complaints will gradually be dispelled by the waters from the springs" (Rindl, 1936: 1). The great benefits of a healthy climate being linked to medicinal springs were re-stated. It was stressed that South Africa "is recognised as possessing one of the most enjoyable and healthful climates in the world" and therefore that its medicinal springs·all had "this great advantage – they are 'sunshine' springs" (Rindl, 1936: 5).

To reinforce the point, detailed comparisons were drawn between the numbers of sunshine hours in South Africa as compared to London, Hamburg, Brussels and Vienna. The key message was undisguised that "if the waters of two medicinal springs were identical as regards chemical composition, temperature etc., but if one of the springs were situated in a locality the climate of which was dull, damp and depressing, whereas the other was in a pleasant locality with a bright sunny climate the 'sunshine' spring would have a greater curative value" (Rindl, 1936: 5). For potential visitors it was argued that the advantages of South African sunshine outweighed the fact that the "spas of a young and thinly-populated country cannot expect to vie with those of Europe and America, which offer all the facilities for treatment which medical science can devise, and which are equipped with every refinement of modern luxury" (Rindl, 1936: 1). A national guide to 'sunshine' railway travel in South Africa aimed at overseas tourists and produced in 1937 included sections on the spas of Caledon described as "medicinal springs of first importance", Warmbaths with its "health-giving waters" and Montagu where" the baths have held the reputation of possessing great curative properties" (Carlyle-Gall, 1937). Again, comparisons were drawn with the spas in Europe; at Montagu the message was repeated now that the town's "waters are authoritatively ranked as more highly radio-active than those of Karlsbad and Baden-Baden" (Carlyle-Gall, 1937: 96).

The above promotional messaging about medicinal springs continued in a 1952 pamphlet issued by South African Railways (Kent, 1952). The benefits of climate in relation to its medicinal springs were reiterated: "Sunshine and pleasant surroundings unquestionably aid the action of spa waters" (Kent, 1952: 1). The range in the chemical composition of the country's medicinal springs was highlighted from alkaline springs, salt springs to sulphurated waters and chalybeatic waters. Importantly attention was drawn to the fact that at "nearly all our South African medicinal springs the water is naturally warm or hot and is not artificially heated as is the case at some of the best known European and American spas" (Kent, 1952: 1). A further advantage of spas in South Africa was that "at many places one may bathe in the open air in the actual 'eye' of the spring, where the water is still gassy and the desirable property of 'freshness,' stressed by so many balneotherapists, has not been lost" (Kent, 1952: 1). A caution was offered, however, that "the visitor to a South African spa, even one of the better equipped establishments, will not find there the ornate casinos, opera houses and pavilions that are deemed essential adjuncts to the leading spas of Europe" (Kent, 1952: 1). Rather, overseas visitors were directed to appreciate the "unwonted charm in the simplicity and naturalness" of South Africa's less sophisticated small town spa destinations (Kent, 1952: 1).

10.4 Caledon – Historical Tourism Pathway of a Small Spa Town

The different individual historical pathways of development of South Africa's spa resorts, which in some instances involves the rise and fall of local tourism, remain mostly undocumented in small town tourism scholarship. This section offers a contribution with a local study of one small town Caledon which is located in the Western Cape 87 miles from Cape Town (Fig. 10.1). The discussion on this small town draws from documentary sources sourced at the National Library depot (Cape

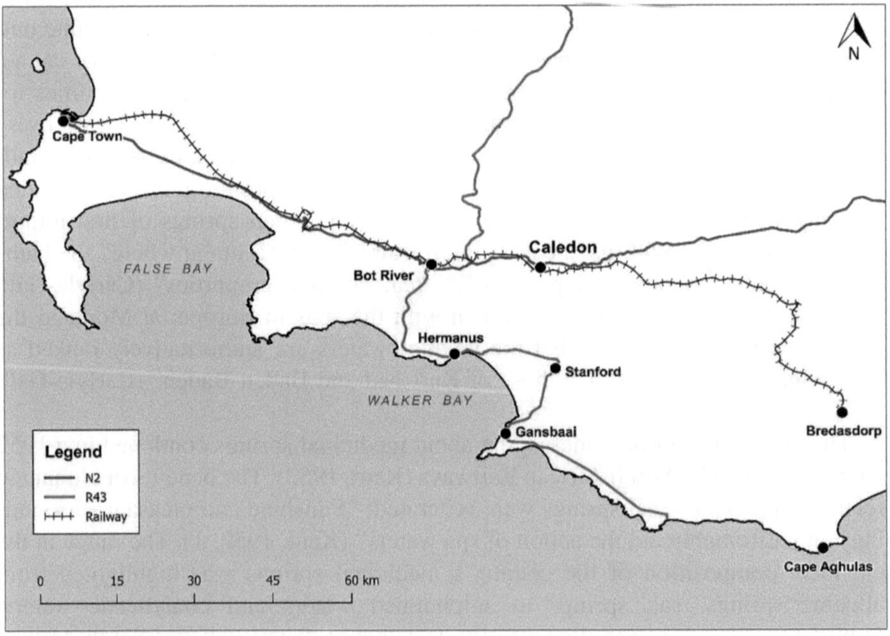

Fig. 10.1 Location map of Caledon. (Source: Authors)

Town), the Historical Papers of the University of the Witwatersrand (Johannesburg) as well as the account of the town's social history given by van Wyk (2013).

The Caledon hot mineral springs have a long recorded history of usage with first accounts going back to the period of Dutch settlement albeit with recognition of earlier use by indigenous populations (van Wyk, 2013). Kent (1952: 2) stated that "the natives took the waters chiefly as mud-baths and that they were highly efficacious in the cure of cutaneous, rheumatic and other diseases as well as for sufferers of pulmonary complaints". Arguably the iron-rich springs featured in an important way in the birth (1811) and development of this small town" (van Wyk, 2013). The government of the Cape Colony recognised that the Caledon bath was an asset and encouraged that it be turned into a resort that resembled European spas making it an attraction for the Colony. The period of the nineteenth century witnessed an increasing popularity of Caledon as a place to take the waters and relative to other South African thermal spas Caledon benefited from its close proximity to Cape Town. Boekstein (2012: 31) records that initially "the spring was utilized only by those seeking some form of medical treatment, but when a new bathhouse, with furnished rooms and excellent food was completed in 1821, the baths became popular with people from all walks of life, for pure relaxation as well as for medicinal purposes". By the close of the nineteenth century the credentials of Caledon as a health resort based upon the properties of its mineral springs were well-established. A report on health resorts which was published in the influential *British Medical Journal* in 1897 observed that the "picturesque village of Caledon is beautifully situated on the southern slopes of a range of mountains, about 830 feet above sea level, about two hours by rail and five hours by cart from Cape Town" (Danielle, 1897: 407). Its potential value for medical practitioners was asserted: "Caledon can be used as a health resort all the year round" (Danielle, 1897: 407).

The most critical development was the upgrading of the bath facilities to establish a Sanatorium in 1897 with the subsequent enhancement of accommodation facilities in terms of a hotel (van Wyk, 2013). This process was driven by two businessmen from Cape Town, the Walsh brothers who implemented plans to turn the old baths into a world class Sanatorium. Although the initial focus was improvement and modernisation of the bath facilities there was an acknowledgment that "the amount of accommodation as well as comfort had to be drastically improved so that the visitors to the baths would have access to the best they could offer" (van Wyk, 2013: 125). By 1897 the treatments offered at Caledon had been upgraded and modernised with the use of modern medical technologies in terms of its treatment bathrooms (Danielle, 1897). The Sanatorium itself was completed in 1897 designed to rival the spas of Europe but the quality of available accommodation required further enhancement. As described by van Wyk (2013: 127) the two-storey sanatorium was "a masterpiece and feast for the eye" and resembled a Swiss Chalet with the illusion of a visit to a European spa (Fig. 10.2).

The building was a T-shaped one with wooden verandah floors and there was a beautiful view from each of the balconies, which gave patients shelter from the elements like wind and rain. The dining room has stunning wide glazed French doors and [a] 120 people could enjoy dinner in the room. Keeping with the Victorian feel,

Fig. 10.2 Caledon Sanatorium and hotel. (Source: Union Publishing Agency, 1924)

there was a reading and writing room, as well as a ladies drawing room that opened up on the veranda. The ground floor had twenty-five bedrooms, a consultation room for the doctor, a billiard room and smoking room and the office of the manager. Bathrooms which were fitted with treatment baths could also be found on this floor. Six in total and were situated so that the minimum distance had to be covered by a patient between the bathroom and his room (van Wyk, 2013: 128).

In 1899 the local municipality produced a brochure which proclaimed that Caledon with its spa was "the premier health resort of the Cape" (Caledon Municipality, 1899). Another key moment was the 1902 completion of the direct railway link with Cape Town which shortened travel times. A 1907 guidebook on the Cape Colony produced by the Cape Government railway department flagged Caledon's emergence as a tourist centre linked to the local asset of the thermal chalybeate spring which was described as "the most remarkable spring in South Africa". Burton (1907: 52) describes it as follows: "During the past few years Caledon has become famous among tourists, both on account of the attractiveness of its scenery and its virtues as a health resort". Although Caledon was never ranked among South Africa's leading health resorts for tuberculosis sufferers (Rogerson & Rogerson, 2021e) "the open air treatment for consumptives has been recommended by medical practitioners, the climate being dry, healthy and bracing, and as all modern conveniences are procurable locally, especially since the railway was extended to the town" (Burton, 1907: 52). The curative value of the baths for other ailments was stressed: "It is seldom that a course of treatment for rheumatism, gout, kidney diseases or lumbago, fails to give quick relief at the Caledon chalybeate baths: indeed, in many cases complete cures have been effected" (Burton, 1907: 52).

Beyond its assets for early health tourism the Caledon surrounds was attractive for hunters. It was noted that: "Sportsmen from all parts of the Colony visit the district during the unprotected game season, for antelopes, pheasants, partridges, hares, snipe, duck and other wild fowl" (Burton, 1907: 52). Caledon was consolidating as a health and holiday resort and with its accessibility enhanced by the rail link an influx of tourists was anticipated with the small town fitting a gap in the market for a suitable winter and summer resort in South Africa. Further new additions were made including a new wing with a dancing and concert hall as well as a lounge. The European feel of the Sanatorium was reinforced by the importation of building materials such as ceramic tiles and cast iron railings (van Wyk, 2013). This "was in the stark contrast to the historic 'primitive Spa' that Caledon baths once were" (van Wyk, 2013: 161). With its added nature attractions of flora and hunting Caledon became marketed to visitors more widely in guidebooks and tourist publications with its anchor being the Sanatorium to make it a small town spa to rival those in Europe. The Sanatorium was finished in 1904 and with its new developments was targeted to "not only draw South African visitors but also European visitors who for a long time lamented the lack of entertainment at the Baths" (van Wyk, 2013: 160). By 1911 the Caledon Baths Ltd. was advertising the Sanatorium hotel and baths as 'nature's gift of gifts' and highlighting its "marvellous cures" in both acute and chronic cases of rheumatism, gout and allied diseases (Cape Times, 1911: 24). Figure 10.3 shows the type of marketing for Caledon Baths which appeared in the journal *Motoring in South Africa* in 1912 and proclaimed the waters as world renowned.

After 1904 the improvements were centred on accommodation such that by the 1920s Caledon could boast a Sanatorium and Hotel which made this small town the "premier holiday and health resort of the Colony" (Union Publishing Agency, 1923). In 1924 the Caledon entry in the *Illustrated South African Hotel Guide*, valued by both domestic and overseas visitors, spoke of this small town "has a beauty all its own, absolute and unique, but to its natural advantages there are the Baths" (Union Publishing Agency, 1924: 43). It was recorded that visitor numbers were increasing annually "and it is doubtful if there is any inland town in Africa, that can compare with Caledon as a health and holiday resort" (Union Publishing Agency, 1924: 43). The Caledon baths were considered "too well known to need any description" and that the hotel "built below the site of the natural spring, is splendidly appointed, and one can spend a most delightful holiday there, for the company is ever congenial; there is golf and – if one knows the ropes – shooting; there are the mountains, and as to the flora, it is matchless – superb" (Union Publishing Agency, 1924: 45). The guide gave considerable play to the natural assets of Caledon as tourism attractions: "The wax heaths of the Caledon Spring Show have drawn visitors from England. The scenery is very like that of some of the glens of Scotland, whilst the famous Blue-bells of Caledon will, when a national poet arises, challenge those of Scotland" (Union Publishing Agency, 1924: 45). Overall it was observed the "road from Cape Town to Caledon is indescribably beautiful, and for those in search of a quiet health-giving holiday, with many new experiences, there can be nothing better than Caledon" (Union Publishing Agency, 1924: 45).

Fig. 10.3 Caledon baths advert. (Source: Union Publishing Agency, 1923)

In 1931 a promotional pamphlet produced by the local Chamber of Commerce to introduce Caledon as a tourist destination gave considerable attention to the springs as well as Caledon's beauty as "the wildflower garden of the Cape" (Caledon Chamber of Commerce, 1931). Indeed, to an increasing extent, during the 1930s Caledon was compared to famous spa towns such as Bath in England "given that Bath also had thermal chalybeate waters and there were very high hopes that Caledon would follow in Bath's footsteps" (van Wyk, 2013: 164). The 1936 national guide to accommodation services in South Africa indicated the Caledon Baths as the largest establishment with two other small hotels and two boarding houses also available (South African Railways and Harbours, 1936). Accommodation at the Baths was the most expensive in town. A 1940 guide described Caledon as a "famous spa and health resort" with "medicinal springs of high curative value"; other assets of the town were listed as "golf, bowls, tennis" (South African Railways and Harbours, 1940: 14). By 1940 the fame of the Sanatorium and its hotel was gaining wide attention and plaudits. Darley-Hartley (1940: 438) draws attention to "superbly fitted" bathrooms, "perfect" sanitary arrangements, the "very fine" concert hall, "a musicians' gallery and a grand piano" and the "food and its serving equal those of a first-class Johannesburg hotel". Above all, the Sanatorium and hotel was "a superb building, containing about one hundred and fifty bedrooms, and every other appurtenance of a first-class *Établissement de Bains* in France or Switzerland" (Darley-Hartley, 1940: 438).

From an historical vantage point the half-century following the building of the Sanatorium represented 'golden years' for the small town of Caledon with its rise and establishment as a spa tourism resort. The year 1946, however, represented a 'turning point' in the economic fortunes of this small town. On the morning of June 1946 a fire started in the dancing hall and the famed Sanatorium burned down beyond recognition. For the tourism economy this was a 'path-changing' moment as the main tourism attraction that had given Caledon its popularity was destroyed with the result an enormous revenue loss for the town. As van Wyk (2013: 265) recorded the town "came to a standstill" as there were no parallel local attractions to continue to draw large numbers of tourists. Following this disaster, the property with what remained of the Sanatorium changed ownership on several occasions with many unfulfilled promises of rebuilding the once glorious building (van Wyk, 2013). The spa town of Caledon literally fell off the tourism map with its core asset no longer. Although no actual data is available on tourism numbers the demise of the tourism economy of Caledon is reflected in the state of local accommodation services. South Africa's first national directory of graded hotels which was produced in 1972 listed only three small one-star accommodation establishments for Caledon; these provided in total only 29 rooms for potential visitors (Hotel Board, 1972).

With the fall of Caledon as a spa destination the residue of the local tourism economy essentially became reliant upon the local attraction of its natural flora, a day trip excursion from the rising coastal resort town of Hermanus (Rogerson & Rogerson, 2020c). Indeed, visiting the wild flower garden of Caledon was advertised as one of a number of 'exciting' day excursions to be enjoyed from a stay in

Hermanus (Hermanus Publicity Association, 1961). In 1980 despite many failed proposals for resuscitating Caledon's core tourism asset, there was still no maximisation of the small town's unique mineral waters. Only in 1990 at the initiative of a local businessman was a project launched for reinstating the curing thermal waters of Caledon as the 'Overberg spa'. This new development, including a 100 room hotel, represented a re-invention of Caledon as a destination within the growing economy of 'wellness tourism'. This was a shift towards a family-oriented resort, a major departure and change to Caledon as the spa tourism destination of the first half of the twentieth century. Another aspect of the reinvention of Caledon tourism was the award to the small town of the first licence for a casino in the Western Cape. With great fanfare the casino with 250 slot machines and 14 gambling tables was opened on 11 October 2000.

10.5 Conclusion

Spas represent a distinct category of leisure resorts with their historical evolution encouraging a small but expanding international literature (Walton, 2014). It has been shown that in several parts of the world the asset of mineral springs has provided a historical springboard for small town tourism development. In South Africa tourism development associated with the country's mineral springs assets never reached the scale of the famous spa resorts of Europe and North America. Beginning in the nineteenth century, however, a number of mineral springs in South Africa became focal points for tourism development and economic progress of small towns. Arguably, both the motives of health and recreation were drivers for the rise of small town spa destinations in South Africa. The individual development paths of these small towns in South Africa inevitably exhibited different local trajectories.

In the case of Caledon this small town established its presence by the 1890s and enjoyed its heyday as a spa tourism destination for the next half-century. The town's dramatic demise as a tourism destination was swift and followed upon the fire that devastated the Sanatorium and the hotel in 1946. For almost 50 years what remained of Caledon's tourism economy was disconnected from this small town's major asset. During the 1990s Caledon tourism experienced some revival with the building of a spa hotel and subsequently the successful bid and award of one of the casino licences for Western Cape province. At one level, this revival of Caledon tourism can be interpreted as an adaptive response to the new milieu of wellness tourism as "even though modern spa treatments were made available, the thermal waters were still honoured as the biggest draw card" for visitors (van Wyk, 2013: 301). Further, the Caledon record underlines the uneven development of spa destinations and the particular fragilities of growing small town tourism in South Africa which is anchored on the asset of mineral springs.

Acknowledgements Thanks to Arno Booyzen who prepared Fig. 1 and to helpful inputs to the preparation of this chapter given by Lulu White as well as Robbie and Skye Norfolk. The University of Johannesburg provided research funding.

References

Alonso-Álvarez, L. (2012). The value of water: The origins and expansion of thermal tourism in Spain, 1750–2010. *Journal of Tourism History, 4*(1), 15–34.

Bell, M. (1993). 'The pestilence that walketh in darkness': Imperial health, gender and images of South Africa c. 1880–1910. *Transactions of the Institute of British Geographers, 18*, 327–341.

Boekstein, M. (2012). *Revitalising the healing tradition – Health tourism potential of thermal springs in the Western Cape.* [Doctoral dissertation]. Cape Peninsula University of Technology.

Boekstein, M. (2014a). Tourism, health and the changing role of health springs – Should South Africa reposition its thermal spring tourism product? *African Journal of Hospitality, Tourism and Leisure, 3*(2), 1–9.

Boekstein, M. (2014b). Healing waters: Balneological classification of thermal springs in South Africa. *African Journal for Physical Health Education, Recreation and Dance, 20*(21), 557–568.

Boekstein, M. (2014c). From illness to wellness – Has thermal spring health tourism reached a new turning point? *African Journal of Hospitality, Tourism and Leisure, 3*(2), 1–11.

Boekstein, M., & Spencer, J. P. (2013). International trends in health tourism: Implications for thermal spring tourism in the Western Cape Province of South Africa. *African Journal for Physical Health Education, Recreation and Dance, 19*(2), 287–298.

Booyens, I. (2020). Tourism innovation in the Western Cape, South Africa: Evidence from wine tourism. In J. M. Rogerson & G. Visser (Eds.), *New directions in South African tourism geographies* (pp. 183–202). Springer.

Borsay, P. (2012). Town or country? British spas and the urban-rural interface. *Journal of Tourism History, 4*(2), 155–169.

Brodie, A. (2012). Scarborough in the 1730s – Spa, sea and sex. *Journal of Tourism History, 4*(2), 125–153.

Burton, A. R. E. (1907). *Cape Colony To-day.* Cape of Good Hope Department of Railways.

Butler, G., & Rogerson, C. M. (2016). Inclusive local tourism development in South Africa: Evidence from Dullstroom. *Local Economy, 31*(1/2), 264–281.

Caledon Chamber of Commerce. (1931). *Caledon and its district: The wildflower garden of the Cape.* South African Railways and Harbours.

Caledon Municipality. (1899). *Caledon and its spa: The premier health resort of the Cape, South Africa.* Dennis Edwards.

Cape Times. (1911). *The South African hotel guide: "Where to stay in South Africa".* Cape Times Ltd..

Carlyle-Gall, C. (1937). *Six thousand miles of sunshine travel over the South African Railways.* South African Railways and Harbours.

Cross, G. (2012). Saratoga Springs: From genteel spa to Disneyfied family resort. *Journal of Tourism History, 4*(1), 75–84.

Danielle, G. W. R. (1897). Notes on health resorts and sanatoria. *The British Medical Journal*, 14 April, 407–408.

Darley-Hartley, W. (1940). The mineral waters of Caledon. *South African Medical Journal,* 23 November, 437–442.

Davidson, J. (2000). Coping with marginality: Tourism and the projection of Grahamstown, 1870–1955. *South African Historical Journal, 42*, 176–190.

Dlomo, T., & Rogerson, C. M. (2021). Tourism and local economic development in King Sabata Dalindyebo Municipality. *African Journal of Hospitality, Tourism and Leisure, 10*(1), 145–164.

Donaldson, R. (2007). Tourism in small town South Africa. In C. M. Rogerson & G. Visser (Eds.), *Urban tourism in the developing world: The South African experience* (pp. 307–325). Transaction Publishers.

Donaldson, R. (2009). The making of a tourism gentrified town: Greyton, South Africa. *Geography, 94*(2), 88–99.

Donaldson, R. (2018). *Small town tourism in South Africa*. Springer.

Donaldson, R. (2021). Small town tourism in South Africa revisited. In C. M. Rogerson & J. M. Rogerson (Eds.), *Urban tourism in the global South: South African perspectives* (pp. 193–211). Springer Nature.

Drummond, F. (2021). The role of tourism in small town cultural and creative industries clustering: The Sarah Baartman District, South Africa. In C. M. Rogerson & J. M. Rogerson (Eds.), *Urban tourism in the global South: South African perspectives* (pp. 213–237). Springer Nature.

Drummond, J., Drummond, F., & Rogerson, C. M. (2021). Latent opportunities for heritage tourism in South Africa: Evidence from Mahikeng and surrounds. *African Journal of Hospitality, Tourism and Leisure, 10*(5), 1591–1609.

Drummond, J., Rogerson, C. M., & Drummond, F. (2022). Adventure tourism in the apartheid era: Skydiving in Mafikeng-Mmabatho. *African Journal of Hospitality, Tourism and Leisure, 11*(SE1), 578–594.

Foster, J. (2003). 'Land of contrasts' or 'home we have always known': The SAR&H and the imaginary geography of white South African nationhood, 1910–1930. *Journal of Southern African Studies, 29*(3), 657–680.

Frost, W., & Laing, J. (2016). History of spa tourism: Spirituality, rejuvenation and socialisation. In M. K. Smith & L. Puczko (Eds.), *The Routledge handbook of health tourism* (pp. 9–19). Routledge.

Giddy, J. K., & Rogerson, J. M. (2021). Nature-based tourism enterprise adaptive responses to COVID-19 in South Africa. *GeoJournal of Tourism and Geosites, 36*(2 Supplement), 698–707.

Giddy, J. K., Rogerson, C. M., & Rogerson, J. M. (2022). Rural tourism firms in the COVID-19 environment: South African challenges. *GeoJournal of Tourism and Geosites, 41*(2), 343–353.

Gordon, B. M. (2012). Reinventions of a spa town: The unique case of Vichy. *Journal of Tourism History, 4*(1), 35–55.

Hembry, P. (1990). *The English spa 1560–1815: A social history*. Athlone Press.

Hermanus Publicity Association. (1961). *Hermanus: The Riviera of South Africa*. Publicity Association.

Hotel Board. (1972). *Guide to the Hotels in South Africa*. Chris van Rensburg Publications.

Houghton, J., Dlamini, B., & Methembu, N. (2013). Considering small towns research and local economic development in South Africa. *Skills at Work: Theory and Practice Journal, 6*, 13–25.

Jacobs, L., du Preez, E., & Fairer-Wessels, F. (2020). To wish upon a star: Exploring Astro-tourism as a vehicle for sustainable rural development. *Development Southern Africa, 37*(1), 87–104.

Johnson, R. M. (2013). *Spas and seaside resorts in Kent, 1660–1820* [Doctoral Dissertation]. University of Leeds.

Kent, L. (1952). *The Medicinal Springs of South Africa*. South African Railways.

Kontsiwe, N., & Visser, G. (2019). Tourism as a vehicle for local economic development in small towns? When things go wrong: The case of Aliwal North, South Africa. *GeoJournal of Tourism and Geosites, 27*(4), 1334–1346.

Lourens, M. (2007). Route tourism: A roadmap for successful destinations and local economic development. *Development Southern Africa, 24*(3), 475–490.

McKay, T. (2013). Development and land use conflicts on the Ash River, South Africa: Energy provision versus adventure tourism. *African Journal for Physical, Health Education, Recreation and Dance, 19*(Supplement 2), 276–293.

McKay, T. (2020). Locating great white shark tourism in Gansbaai, South Africa within the global shark tourism economy. In J. M. Rogerson & G. Visser (Eds.), *New directions in South African tourism geographies* (pp. 283–296). Springer.

McMorran, C. (2008). Understanding the 'heritage' in heritage tourism: Ideological tool or economic tool for a Japanese hot springs resort? *Tourism Geographies, 10*(3), 334–354.

Norval, A. J. (1936). *The tourist industry: A national and international survey*. Sir Isaac Pitman.

Pirie, G. H. (2011). Elite exocitism: Sea-rail cruise tourism to South Africa, 1926–1939. *African Historical Review, 43*(1), 73–99.

Rindl, M. (1936). *The Medicinal Springs of South Africa*. South African Railways and Harbours.

Rogerson, C. M. (2007). Tourism routes as vehicles for local economic development in South Africa: The example of the Magaliesberg Meander. *Urban Forum, 18*(2), 49–68.

Rogerson, C. M. (2014). Reframing place-based economic development in South Africa: The example of local economic development. *Bulletin of Geography: Socio-Economic Series, 24*, 203–218.

Rogerson, J. M. (2019). The evolution of accommodation services in a coastal resort town: Hermanus, South Africa. *African Journal of Hospitality, Tourism and Leisure, 8*(5), 1–16.

Rogerson, C. M. (2020). Using municipal tourism assets for leveraging local economic development in South Africa. *Bulletin of Geography: Socio-Economic Series, 48*, 47–63.

Rogerson, J. M. (2021). Tourism business responses to South Africa's COVID-19 state of emergency. *GeoJournal of Tourism and Geosites, 35*(2), 338–347.

Rogerson, C. M., & Rogerson, J. M. (2014). Agritourism and local economic development in South Africa. *Bulletin of Geography: Socio-Economic Series, 26*, 93–106.

Rogerson, C. M., & Rogerson, J. M. (2019). Tourism, local economic development and inclusion: Evidence from Overstrand Local Municipality, South Africa. *GeoJournal of Tourism and Geosites, 25*(2), 293–308.

Rogerson, C. M., & Rogerson, J. M. (2020a). Inclusive tourism and municipal assets: Evidence from Overstrand Local Municipality, South Africa. *Development Southern Africa, 37*(5), 840–854.

Rogerson, C. M., & Rogerson, J. M. (2020b). Municipal caravan parks in South Africa: Geography, management and planning. *Bulletin of Geography: Socio-Economic Series, 49*, 109–124.

Rogerson, C. M., & Rogerson, J. M. (2020c). Resort development and pathways in South Africa: Hermanus, 1890–1994. In J. M. Rogerson & G. Visser (Eds.), *New directions in South African tourism geographies* (pp. 15–32). Springer.

Rogerson, C. M., & Rogerson, J. M. (2021a). The other half of urban tourism: Research directions in the global south. In C. M. Rogerson & J. M. Rogerson (Eds.), *Urban tourism in the global south: South African perspectives* (pp. 1–37). Cham, Switzerland.

Rogerson, C. M., & Rogerson, J. M. (2021b). Municipal assets for tourism and small-town local economic development: Caravan parks in South Africa. In J. Saarinen & J. M. Rogerson (Eds.), *Tourism, change and the global south* (pp. 205–221). Routledge.

Rogerson, C. M., & Rogerson, J. M. (2021c). Creative networks and the making of Africa's first UNESCO creative city of gastronomy. In C. M. Rogerson & J. M. Rogerson (Eds.), *Urban tourism in the global south: South African perspectives* (pp. 239–266). Cham, Switzerland.

Rogerson, C. M., & Rogerson, J. M. (2021d). Mundane urban tourism: The historical evolution of caravan parks in South Africa 1930–1994. In C. M. Rogerson & J. M. Rogerson (Eds.), *Urban tourism in the global South: South African perspectives* (pp. 93–112). Springer Nature.

Rogerson, C. M., & Rogerson, J. M. (2021e). Climate therapy and the development of South Africa as a health resort, c1850–1910. *Bulletin of Geography: Socio-Economic Series, 52*, 111–121.

Rogerson, C. M., & Rogerson, J. M. (2022a). The first round impacts of COVID-19 for rural tourism in South Africa. *Studia Periegetica, 37*(1), 63–86.

Rogerson, C. M., & Rogerson, J. M. (2022b). COVID-19 and coastal destination impacts: The case of South Africa. *GeoJournal of Tourism and Geosites, 42*(2), 767–774.

Rogerson, J. M., & Visser, G. (2020). Recent trends in south African tourism geographies. In J. M. Rogerson & G. Visser (Eds.), *New directions in south African tourism geographies* (pp. 1–13). Springer.
Rogerson, J. M., Lekgau, R., Mashapa, M., & Rogerson, C. M. (2021a). COVID-19 and local business responses: Evidence from South Africa's most tourism-dependent locality. *African Journal of Hospitality, Tourism and Leisure, 10*(1), 388–405.
Rogerson, J. M., Lekgau, R., Mashapa, M., & Rogerson, C. M. (2021b). Rural tourism challenges in South Africa under COVID-19: Greater Giyani Local Municipality, Limpopo Province. *African Journal of Hospitality, Tourism and Leisure, 11*(3), 1099–1117.
Rotherham, I. D. (2014). *Spas and spa visiting*. Bloomsbury for Shire Publications.
Smith, M. K. (2008). Regeneration of an historic spa town: A case study of Spa in Belgium. In M. Smith & L. Puczko (Eds.), *Health and wellness tourism* (pp. 300–304). Routledge.
South African Railways and Harbours. (1928). *The Medicinal Springs of South Africa*. South African Railways and Harbours.
South African Railways and Harbours. (1930). *Aliwal North and its wonder pools*. South African Railways and Harbours.
South African Railways and Harbours. (1936). *Illustrated guide to hotels and boarding houses*. South African Railways and Harbours.
South African Railways and Harbours. (1940). *Illustrated guide to hotels, boarding houses and farm guest houses*. South African Railways and Harbours.
Symes, T. E. (1889). *South Africa as a health resort*. Searle & Rivington.
Union Publishing Agency. (1923). *The illustrated South African hotel guide, 1923*. Cape Times Limited.
Union Publishing Agency. (1924). *The illustrated South African hotel guide, 1924*. Cape Times Limited.
Van Wyk, D. (2013). *The social history of three Western Cape thermal mineral springs resorts and their influence on the development of the health and wellness tourism industry in South Africa* [Masters dissertation]. Stellenbosch University.
Visser, G. (2016). South African tourism geographies: Progress and prospects. *South African Geographical Journal, 98*(3), 428–438.
Walton, J. K. (1983). *The English seaside resort: A social history, 1750–1914*. Leicester University Press.
Walton, J. K. (2009). Histories of tourism. In T. Jamal & M. Robinson (Eds.), *The SAGE handbook of tourism studies* (pp. 115–129). SAGE.
Walton, J. K. (2011). The history of British spa resorts: An exceptional case in Europe? *Transportes, Servicios y Telecomunicaciones, 20*, 139–157.
Walton, J. K. (2012). Health, sociability, politics and culture. Spas in history, spas and history: An overview. *Journal of Tourism History, 4*(1), 1–14.
Walton, J. K. (2014). *Mineral springs resorts in global perspective: Spa Histories*. Routledge.
Wessels, W., & Tseane-Gumbi, L. (2022). Identifying urban tourism development challenges in Mahikeng. *African Journal of Hospitality, Tourism and Leisure, 11*(1), 14–25.
White, R. (2012). From the majestic to the mundane: Democracy, sophistication and history among the mineral spas of Australia. *Journal of Tourism History, 4*(1), 85–108.

MIX
Papier aus verantwortungsvollen Quellen
Paper from responsible sources
FSC® C105338
FSC
www.fsc.org

If you have any concerns about our products,
you can contact us on
ProductSafety@springernature.com

In case Publisher is established outside the EU,
the EU authorized representative is:
Springer Nature Customer Service Center GmbH
Europaplatz 3, 69115 Heidelberg, Germany

Printed by Libri Plureos GmbH
in Hamburg, Germany